SMALL MISSIONS FOR ENERGETIC ASTROPHYSICS

SMALL MISSIONS FOR ENERGETIC ASTROPHYSICS

Ultraviolet to Gamma-Ray

Los Alamos, New Mexico   February 1999

EDITOR
Steven P. Brumby
Los Alamos National Laboratory

AIP CONFERENCE PROCEEDINGS 499

American Institute of Physics   Melville, New York

Editor:

Steven P. Brumby
Space and Remote Sensing Sciences
Los Alamos National Laboratory
Mail Stop D436
Los Alamos, NM 87545
USA

E-mail: brumby@nis.lanl.gov

Authorization to photocopy items for internal or personal use, beyond the free copying permitted under the 1978 U.S. Copyright Law (see statement below), is granted by the American Institute of Physics for users registered with the Copyright Clearance Center (CCC) Transactional Reporting Service, provided that the base fee of $15.00 per copy is paid directly to CCC, 222 Rosewood Drive, Danvers, MA 01923. For those organizations that have been granted a photocopy license by CCC, a separate system of payment has been arranged. The fee code for users of the Transactional Reporting Service is: 1-56396-912-2/99/$15.00.

© 1999 American Institute of Physics

Individual readers of this volume and nonprofit libraries, acting for them, are permitted to make fair use of the material in it, such as copying an article for use in teaching or research. Permission is granted to quote from this volume in scientific work with the customary acknowledgment of the source. To reprint a figure, table, or other excerpt requires the consent of one of the original authors and notification to AIP. Republication or systematic or multiple reproduction of any material in this volume is permitted only under license from AIP. Address inquiries to Office of Rights and Permissions, Suite 1NO1, 2 Huntington Quadrangle, Melville, N.Y. 11747-4502; phone: 516-576-2268; fax: 516-576-2450; e-mail: rights@aip.org.

L.C. Catalog Card No. 99-068682
ISBN 1-56396-912-2
ISSN 0094-243X
Printed in the United States of America

Contents

Preface vii
Committees ix
Small Missions for Energetic Astrophysics: A Panel Discussion 1
    S. P. Brumby and W. C. Priedhorsky
Soft X-Ray, EUV and Far-UV Studies of Hot White Dwarfs 8
    M. A. Barstow
Wide-Field All-Sky Monitor for X-Ray Astronomy 20
    K. N. Borozdin, W. C. Priedhorsky, V. A. Arefiev,
    A. S. Kaniovsky, K. Black, and S. Brandt
BALLERINA—Pirouettes in Search of Gamma Burst Sources 33
    S. Brandt and N. Lund
High Resolution X-Ray Imaging 44
    W. Cash
The Intergalactic Medium and Soft X-Ray Background 58
    R. Cen
Perspectives of Astrophysical and Gravitational Research Onboard Small Spacecraft 68
    V. I. Denisov, S. I. Svertilov, M. I. Kudryavtsev,
    Z. P. Cheryomukhina, and V. B. Pinchoock
Neutron Starquakes 75
    L. M. Franco, R. I. Epstein, and B. Link
The ROTSE Detection of Early Optical Light from GRB 990123 82
    G. R. Gisler, C. W. Akerlof, R. J. Balsano, J. J. Bloch,
    D. E. Casperson, S. J. Fletcher, J. G. Hills, R. L. Kehoe,
    B. C. Lee, S. L. Marshall, T. A. McKay, R. S. Miller,
    W. C. Priedhorsky, J. J. Szymanski, and J. A. Wren
The Intergalactic Medium 90
    R. C. Henry
The Interstellar Medium 100
    R. C. Henry
SIXE: An X-Ray Experiment for a Minisatellite 110
    J. Isern, E. Bravo, J. Gómez-Gomar, M. Hernanz,
    E. García-Berro, F. Giovannelli, C. D. La Padula,
    L. Sabau, J. Gutiérrez, J. José, D. García-Senz, J. Bausells,
    J. Cabestany, J. Madrenas, M. Angulo, M. Fernández-Valbuena,
    E. Herrera, M. Reina, and A. Talavera
A Silicon-Strip Coded-Mask Imager for 3–40 keV X-Rays 123
    D. A. Leahy
Discovery Space for Stellar Astrophysics by Small Missions 127
    J. L. Linsky
Instrumentation for a Next-Generation X-Ray All-Sky Monitor 135
    A. G. Peele
Supernova Remnant 146
    H. Tsunemi

CATSAT: A Small Satellite for Studying Gamma-Ray Bursts 156
    W. T. Vestrand, D. J. Forrest, K. A. Levenson, C. Whitford,
    D. Fletcher-Holmes, A. Wells, and A. Owens

Author Index 167

PREFACE

For the next decade, the major milestones for energetic space astrophysics have become clear: FUSE in the ultraviolet, CONSTELLATION in the x-ray, and GLAST in the gamma-ray band. These missions aim squarely at the most important questions of our day. However, small missions (MIDEX, SMEX, UNEX, and others) are also critical. They answer questions unanswered by large missions, particularly because they allow the rapid application of new technologies. Our workshop on "Small Missions for Energetic Astrophysics" brought the community together in a forum to discuss small missions targeted at energetic astrophysics that can be launched in the second half of the next decade.

The workshop was held at the J. Robert Oppenheimer Study Center at Los Alamos National Laboratory, in February of 1999, and attracted representatives from around the world. Approximately 70 scientists attended, including representatives from Universities, NASA Centers, European space agencies, U.S. National and Defense Laboratories, and industry. Major additional funding for the workshop was provided by NASA, and by the Center for Space Science and Exploration and the Institute of Geophysics and Planetary Physics at Los Alamos National Laboratory, which is a U.S. Department of Energy Laboratory operated by the University of California.

Steven P. Brumby
Los Alamos, New Mexico

SCIENTIFIC ORGANIZING COMMITTEE

William C. Priedhorsky, Los Alamos National Laboratory (Chair)
David N. Burrows, Pennsylvania State University
France A. Córdova, University of California at Santa Barbara
George W. Fraser, Leicester University
Neil Gehrels, Goddard Space Flight Center
Jonathan E. Grindlay, Harvard/SAO
Richard C. Henry, Johns Hopkins University
Kevin Hurley, University of California at Berkeley
Mark Hurwitz, University of California at Berkeley
Steve M. Kahn, Columbia University
Nobu Kawai, RIKEN/Tokyo
Don Q. Lamb, University of Chicago
Niels Lund, Danish Space Research Institute
Igor Mitrofanov, IKI/Moscow
Toshio Murakami, ISAS
Arvind Parmar, ESTEC
Wilt T. Sanders, University of Wisconsin at Madison
Nick E. White, Goddard Space Flight Center

LOCAL ORGANIZING COMMITTEE

Steven P. Brumby (Chair)
William C. Priedhorsky
Jeffrey J. Bloch
Bradley C. Edwards
Richard I. Epstein
Edward E. Fenimore
Galen R. Gisler
Cheng Ho
Todd J. Haines
Barham W. Smith

Small missions for energetic astrophysics: A panel discussion

S. P. Brumby and W.C. Priedhorsky

*Space and Remote Sensing Sciences,
Los Alamos National Laboratory, Mail Stop D436,
Los Alamos, NM 87545.*

Abstract. We present the conclusions of a panel discussion held on the final day of the workshop. The major themes discussed by the panel involved increasing standardization, sharing information, improving access to space, and the changing role of the Principal Investigator. We have not attempted to construct a verbatim transcript of the event, but will rather present a summary of the debate and draw some conclusions.

INTRODUCTION

We present the conclusions of a panel discussion held on the final day of the workshop. The panel discussion was intended to address science targets and enabling technologies for small energetic astrophysics missions in the opening decades of the next century. The discussion took the form of a "panel of the whole", with all attendants encouraged to comment throughout the discussion. Approximately 30 scientists attended the discussion, including representatives from Universities, NASA Centers, National and Defense Laboratories, and industry. The session was chaired by William C. Priedhorsky, Chair of the Scientific Organizing Committee.

This report was prepared from an audio recording of the discussion, and from notes taken during and immediately after the session. We have not attempted to construct a verbatim transcript of the event, but will rather present a summary of the debate and draw some conclusions. In the best tradition of scientific workshops, the debate was wide-ranging and quite lively. The panel did not reach full consensus on every point, which we feel helps support our claim that the discussion touched on some of the more important issues presently confronting the small missions (instruments and spacecraft) community.

As with all reports on the deliberations of many, the authors apologize in advance for any misrepresentation of opinion that may occur in what follows.

THE PANEL DISCUSSION

The major themes discussed by the panel involved increasing standardization, improving access to space, and the changing role of the Principal Investigator (PI). We now present these topics in turn.

Standardization

It may seem somewhat paradoxical to attempt to encourage innovative small missions by insisting on greater levels of standardization, but this idea was raised early in the discussion and debated at length. One key problem facing the designer of a small mission is the decision of how much of the instrumentation and support hardware needs to be designed from scratch, how much can be adapted from past missions, and how much can be obtained "off the shelf" (OTS).
Clearly, designing the entire spacecraft and suite of instruments from a blank page may yield an elegant solution, but the design costs will very likely be prohibitive in the context of, e.g., NASA University Explorer (UNEX, of order 10M$), Small Explorer (SMEX; 75M$), and Medium Explorer (MIDEX; 150M$) missions. Adapting an existing spacecraft design might end-up being almost as expensive as a new design, as significant changes to an established design may call into question, if not entirely invalidate, existing engineering models, requiring a substantial pre-flight design validation effort. As for the design of individual instruments, custom interfaces between spacecraft bus and payload instruments ties an instrument to a specific mission, and complicates reuse of designs.

The low-cost alternative is to use generic parts almost everywhere, and seek to innovate in the design and fabrication of the essentially new components, i.e., the payload instruments themselves. This approach to assembling missions has been called a "catalog" approach to space hardware. In fact, NASA maintains a catalog of spacecraft buses through its Rapid Space Development Office (RSDO) [1]. Vendors from the aerospace industry offer configurable buses, and help with the choice of options to meet the needs of the mission planners. These buses come fully equipped with integrated command and communication gear, and important subsystems such as attitude control and power generation. NASA has made available a mission planning and design team to assist proposal writers through its Integrated Mission Design Center (IMDC) [2].

Such an approach clearly goes a long way towards the goal of freeing the mission PI to concentrate on achieving their science goals. Several panelists, however, raised the question of taking this system to its logical conclusion. By analogy to the computer hardware industry, one could consider a future in which industrial players focussed their activities into niche markets. Aerospace companies, supporting technology companies, and new technology start-ups would directly compete to provide each major component of the spacecraft. This competition would be facilitated and protected by the introduction of a set of open standards for spacecraft hardware.

The catalog approach to spacecraft could, in principle, be extended to software. Software development is a critical but sometimes undervalued element of space missions. Software errors can and do cause catastrophic damage to missions, and the total cost of programming and debugging software can rival the cost of the spacecraft bus. Yet software development has a tendency to be under budgeted and under supported relative to the hardware development tasks. For example, one panelist pointed out the extraordinarily high cost of chip-set emulation tools for space-rated embedded computing systems. Such emulators are needed for debugging and validation of flight software. Small space missions are particularly threatened by this state of affairs, as they are reliant on embedded processing systems but are unlikely to be able to afford the expensive software development environments and hardware test-stand equipment available to commercial developers.

A catalog approach might begin to address these issues, especially where one could obtain software guaranteed to be compatible with a desirable spacecraft bus, though such software will necessarily be expensive. Significant further savings may be achieved by shifting to commercial off-the-shelf (COTS) software and parts, where the space environment allows their use, as a means to escape the limitations of traditional "mil-spec" hardware environments. Again, the idea was stated that the community needs to seriously consider moving towards an "open standard" architecture in which networks of independent programmers can work towards innovative, high-quality solutions.

Some panelists commented on the dangers of applying the "open standards" philosophy to space missions. Space is an expensive, difficult, and dangerous domain in which in operate. The financial resources and technological experience needed to develop instruments and sub-system suggests that independent ("backyard") inventors of space hardware will face severe challenges. Further, integration of the hardware and software has historically required close cooperation between all parties. It is not at all clear that modern means of communication allows a physically scattered team to match the performance of a similar multi-disciplinary team working under one roof.

Another important area for increased standardization is in the interface of spacecraft to launch vehicle. This will be discussed in the next section.

Access to Space

The range, availability, and cost of launch opportunities remains a topic of great concern to the small missions community. The launch vehicle market remains extremely expensive, and traditional low-cost launch opportunities, such as sounding rockets, have been repeatedly threatened with cancellation. In principle, small science missions can save substantial sums of money by sharing rides into space, and the panel discussed several aspects of this complex of problems.

Of particular concern is the fact that a significant fraction of launch capacity, estimated by one panelist at 30% to 40%, is simply wasted. The situation is no better with regard to Space Shuttle secondary payloads ("gas cans"), and commercial launch opportunities, e.g., on constellations of communications satellites, have frequently proven to be illusory. The International Space Station represents another opportunity, one with the added advantage that missions may evade some of the usual power and mass constraints encountered on regular free-flying spacecraft platforms.

Part of this wastage is due to the bureaucracy of space. Commercial missions shy away from hosting secondary science payloads for reasons of risk minimization, while, for example, flying an instrument on the Shuttle requires addressing the additional safety concerns attending manned spaceflight. What of the remaining unmanned launches, and how to unlock the growing commercial launch sector? Several panelists suggested the need for information exchange and standardization to solve this problem. Creation of an Office to share knowledge of launch opportunities and of instruments looking for rides, and perhaps help broker agreements, would be an important service to the community. NASA does have a program, called "Quick Ride", managed by the RSDO [1], which seeks to meet the need for advertising launch opportunities, though the majority of the workshop panelists expressed no experience of it.

Another possible key to this issue is the need for standardized interfaces to launch vehicles, enabling small missions to be prepared independently and then matched with the first available launch opportunity. Small missions could benefit even more strongly from interfaces that were common across several launch vehicles, freeing the mission from having to go with a unique make or class of rocket. Such developments would help the international market of launch opportunities, and encourage fair competition between old and new providers of launch services. Whether a truly free market is compatible with national security concerns is a matter for continuing debate, but its scientific and commercial justification seems obvious.

An additional aspect of the problems of increasing and improving access to space, drawn out by the panel, was the challenge of meeting the demands of conservative and risk averse space agencies. With even small missions frequently costing well in excess of 100M$, it is hard to fault Agency demands that proposed flight hardware and scientific instrumentation be proven and reliable. Historically, the solution has been to start with technology demonstration missions using sounding rockets or balloons. Balloon flights seem to be experiencing a minor renaissance, thanks to improvements in long-duration balloon flight capabilities. Similarly, there appears to be continuing role for sounding rockets, with recent fluctuations in demand more a function of concern over the survival of the programs, rather than a reflection of dissatisfaction with the technology. As the next step beyond sub-orbital tests, one can consider programs along the lines of NASA/JPL's New Millennium Project [3], the European Space Agency's Small Missions for Advanced Research in Technology (SMART) program, and the Department of Defense's Space Test Program [4]. These offer flight validation for breakthrough technologies, as most recently demonstrated, respectively, by the Deep Space 1 [5] and ARGOS [6] missions.

Role of the Principal Investigator

Discussion of simplifying the mundane side of the spacecraft design and launch tasks points to a deeper question. Should the small missions community move to a system in which PI's are required to be instrument designers and not spacecraft designers?

Generic mission opportunities using a highly standardized bus design could be advertised (e.g., low-earth orbit, 3-axis stabilized; or Molniya orbit, anti-sun-pointed spinner), managed by NASA or some other large agency. Researchers would then compete for a limited number of instrument payload slots on the platform, designing to a fixed and pre-defined interface. Additional constraints would be on mass, power, and processing/storage/communication resources.

There are some programs that operate in this fashion. For example, NASA/JPL's Mars Surveyor Program [7] is planning a sequence of orbiter and lander missions, each with a theme, such as climate observation or mineral mapping. This program has enjoyed considerable recent success with its Mars Pathfinder and Mars Global Surveyor missions. The European Space Agency has taken a similar approach with its Mars Express mission [8], and international collaborations, such as the Russian/European/US Spectrum-Roentgen-Gamma mission [9], have planned to launch multi-wavelength astronomical observatories carrying comprehensive suites of instruments. Additionally, the U.S. Department of Defense has for a number of years run a Space Test Program [4], offering rides to specifically defense-related technology demonstrations and science payloads.

The key problem with composite missions arises if such missions provide the only opportunities for rides into space. In particular, while the process of picking themes for generic missions would most likely be open, peer-reviewed, and consensus-based, some researchers would doubtless feel unduly constrained or ignored. Research on less popular questions would be in danger of being perpetually marginalized. Hence, introduction of additional composite mission opportunities would be only a partial solution to the needs of researchers more interested in getting data from their instrument than in planning and executing a complete mission, and may not meet the needs of those interested in keeping research fields alive between large missions.

Of course, it can be argued that "small" missions have already grown in complexity to the point where they require teams of institutions, and grown in cost to the point where significant consensus-building is required for their funding. Such a trend needs to be countered, as intended by NASA's UNEX program of very small satellites, and the continuing interest in ultra-small satellite technologies [10].

CONCLUSIONS

The desire to design and fly new small missions is as strong as ever, and, as related throughout this Workshop, proposals for exciting new missions abound. Our community continues to exist in a period of transition, from forced reliance on old technology and large missions, to the new frontier of "faster, smaller, cheaper" missions. Hopes for new and possibly abundant launch opportunities, and for exploiting radically new technologies, have sharpened people's appetite for space, and new classes of missions (e.g., UNEX, SMEX) provide the funding framework for such missions.

The panel discussion helped identify possibly ways to assist the community achieve its hopes. Most suggestions stressed co-operative strategies; sharing of spacecraft design knowledge, sharing of launch opportunity information, catalog approaches to hardware and software acquisition, and acceptance of open standards as a means to free mission designers to focus on their science goals by better exploiting the existing engineering base. A particular need was identified for more opportunities to demonstrate untested instruments and technologies through special low-cost (sounding rocket, balloons) and/or accepted high-risk (e.g., NMP) missions. Many of these needs have already been noticed, and we wish to encourage space agencies, institutions, and professional organizations to continue to consider innovative strategies for enabling small missions in astrophysics and other fields.

NOTES AND REFERENCES

(1) The Rapid Space Development Office (RSDO) is an office of NASA Goddard Space Flight Center's Flight Programs & Projects Directorate. Further information is available online at http://rsdo.gsfc.nasa.gov

(2) The Integrated Mission Design Center (IMDC) NASA Goddard Space Flight Center's Mission Integration & Planning Division. Further information is available online at http://imdc.gsfc.nasa.gov

(3) NASA Jet Propulsion Laboratory (JPL) manages NASA's New Millennium Project (NMP), on the web at http://nmp.jpl.nasa.gov

(4) The Space Test Program (STP) is managed by the United States Air Force (USAF) through the USAF Space and Missile Systems Center's Test and Evaluation Directorate (SMC/TE). Further information on the program is available online at http://www.te.plk.af.mil

(5) First of the NMP missions, Deep Space 1 recently met all of its mission success criteria, and received NASA's Software of the Year award for one of its technologies.

(6) The ARGOS mission, launched in February of 1999, is described on the web at http://www.te.plk.af.mil/arpics/apage1.cfm

(7) NASA/JPL manages NASA's Mars Surveyor Program. Further information on the program is available online at http://mars.jpl.nasa.gov

(8) The European Space Agency's Mars Express is on the web at http://sci.esa.int/marsexpress/

(9) Spectrum-Roentgen-Gamma is on the web at http://hea-www.harvard.edu/SXG/sxg.shtml

(10) The Annual AIAA/Utah State University Conferences on Small Satellites, see http://www.sdl.usu.edu/conferences/index.html

Soft X-ray, EUV and far-UV Studies of Hot White Dwarfs

Martin A. Barstow

Department of Physics and Astronomy
University of Leicester
University Road
Leicester LE1 7RH, UK

Abstract.
   The systematic study of the properties of the structure and evolution of hot white dwarfs owes a tremendous debt to space borne astronomy. At the temperature and densities found in these stars, observations in visible light generally only reveal evidence of the most abundant photospheric elements, hydrogen or helium. Access to spectral windows ranging from the far UV, through the EUV and into the soft X-ray band reveals a much more complex picture of trace heavy element opacity in the stellar atmospheres and interstellar absorption than could ever have been deduced from ground-based measurements alone. Moreover, many of the spacecraft responsible for these results would fall within the 'Small Mission' category. This review discusses the legacy of *IUE*, the *ROSAT* Wide Field Camera and *EUVE*, among others, outlining the need for small and medium Explorer Class missions.

INTRODUCTION

   White dwarfs are the final evolutionary stage of all stars with initial masses below $\approx 7 - 8 M_\odot$, and hence represent the fate of most stars in the Universe. Once a typical $1 M_\odot$ star has ceased buring hydrogen in its core, it passes through an H shell buring red-giant phase, becomes a core He burning subgiant, then a red giant again with H and He shell burning, before becoming unstable and throwing off its outer layers to form a planetary nebula. By this stage, all thermonuclear reactions have ceased and the $\approx 0.5 M_\odot$ degenerate stellar core is revealed, which forms the white dwarf. The white dwarf remnant contains the imprint of the prior evolutionary path, revealing material that has been processed within the core of the progenitor star. In addition, without an internal energy source the effective temperature is determined solely by the rate of dissipation of the internal heat store. Hence, the age of a white dwarf can be determined from its present temperature.
   The study of white dwarfs has, potentially, much to tell us about stellar evolution in general, the physics of matter in extreme conditions, besides the history

and age of the galactic disk. However, to extract this information we need to understand the stars themselves. Unfortunately, studies in the visible region of the electromagnetic spectrum only reveal the spectroscopic signatures of the dominant atmospheric constituents, hydrogen or helium, but do allow the basic identification of two groups of white dwarfs. Most hot white dwarfs ($\approx 80-90\%$) have envelopes which appear to be more or less pure hydrogen. This is not too suprising, as the strong gravitational field of a white dwarf (log g$\approx 7.0-9.0$) will cause the heaviest elements to sink to the bottom of the photosphere leaving a layer of hydrogen on top. The existence of a smaller group of white dwarfs with apparently pure He atmospheres implies that, in some objects at least, almost all the hydrogen has been expelled during earlier phases of mass-loss.

EARLY UV AND EUV/SOFT X-RAY STUDIES

The general division of white dwarfs into H and He-rich types was established before the *IUE* satellite was launched, in 1978. Initial studies of hot white dwarfs anticipated using the featureless UV continuum spectra as background sources to study the effect of interstellar absorption along the line of sight and, as a result, probe the structure of the interstellar medium. It was something of a surprise, therefore, to find absorption features of highly ionized species such as CIV, NV and SiIV in the spectra of several objects (e.g. [9]; see figure 1). However, without knowledge of the stellar radial velocities, the location of this material could not be established, although it was widely proposed to have a circumstellar origin.

The advent of *EXOSAT* in 1983, following on from work with the *Einstein* X-ray

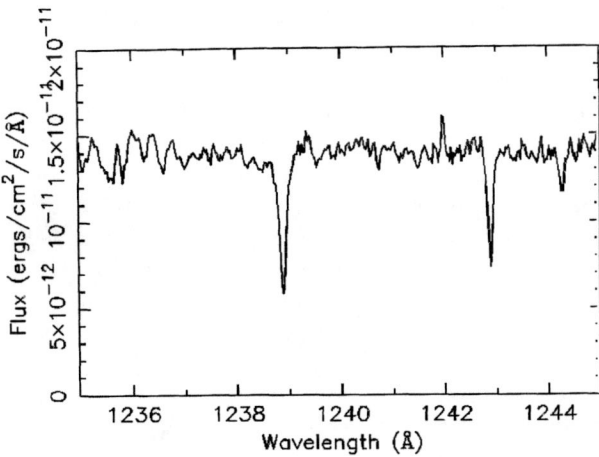

FIGURE 1. NV absorption lines in the *IUE* echelle spectrum of the DA white dwarf G191−B2B.

observatory, added a further dimension to the study of hot white dwarfs, through soft X-ray/EUV photometry. This data clearly revealed significant absorption along the line-of-sight, over and above what could be accounted for by the interstellar medium, which was attributed to the presence of trace helium (He/H$\approx 10^{-5} - 10^{-3}$ e.g. [26]), although there was no direct spectroscopic evidence for the presence of this element in any of the stars studied.

An important contribution was also made by the limited observations performed by the *EXOSAT* tranmission grating spectrometers, before the failure of the grating deployement mechanisms. In all, three white dwarfs were studied, the DA stars HZ43, Sirius B and Feige 24. Both HZ43 and Sirius B revealed the anticipated pure H spectra, with low limits placed on the possible presence of He by the absence of the He Lyman series lines and absorption edge at 228Å [24] [25]. However, the spectrum of Feige 24 had a completely different appearance, being strongly cutoff at short wavelengths when compared with HZ43, a star of similar effective temperature. Vennes et al. [31] explained this with an atmospheric model containing a mixture of elements heavier than either H or He, the first hint that the features discovered by *IUE* might be photospheric rather than circumstellar. Unfortunately, the resolution of the *EXOSAT* spectrometers was too low to have any chance of resolving individual heavy elements and with such a small sample of objects it was not possible to decide whether HZ43 or Feige 24 was representative of the general population of white dwarfs.

THE *ROSAT* ALL-SKY SURVEY

Sources of opacity in white dwarf atmospheres

This background of research underpinned expectations of the results from the first all-sky X-ray and EUV surveys of the sky conducted by *ROSAT*. Based on the belief that photospheric opacities in hot DA white dwarfs were dominated by trace helium, it was predicted that one to two thousand white dwarfs should be detected. In the event, only a small fraction of this number ($\approx 10\%$) were eventually identified in the survey catalogues. A first indication of the possible reason for the shortfall came from examination of the fraction of optically selected PG survey white dwarfs detected. Almost all stars with effective temperatures between 25000K and 40000K are detected but this fraction decreases dramatically at higher temperatures [17]. A detailed study of the observed X-ray and EUV emergent fluxes, after correcting for stellar distance, of all the white dwarfs detected reveals interesting behaviour. Comparing the observed values with the predictions of theoretical stellar model atmosphere calculations shows that stars cooler than 40000K have a more or less pure H composition. However, those stars with temperatures above this exhibit a significant comparative shortfall in the observed fluxes (Figure 2 [1] [23] [20] [33]).

Helium alone cannot explain all the observed opacity and heavier elements must account for a significant fraction of absorption taking place within the stellar photosphere. This proposition is supported by the detection of Fe and Ni ions in *IUE* echelle spectra of the visually brightest white dwarfs showing strong soft X-ray and EUV opacity [18] [19] [32]. This is not in accord with the expectation that all heavy elements should sink out of the envelope in a white dwarf's strong gravitational field. However, theoretical studies indicate that in the hottest white dwarfs radiative forces are large enough to counteract gravity and maintain significant elemental abundances. Species with the largest EUV cross-section (ie. the largest number of transitions) such as Fe and Ni are the most efficiently levitated [12] [13].

White dwarfs in binary systems

A number of the brightest EUV/X-ray sources detected in the *ROSAT* survey were associated with apparently normal stars, but without the apparent level of stellar activity nor the broad band spectral characteristics usually associated with

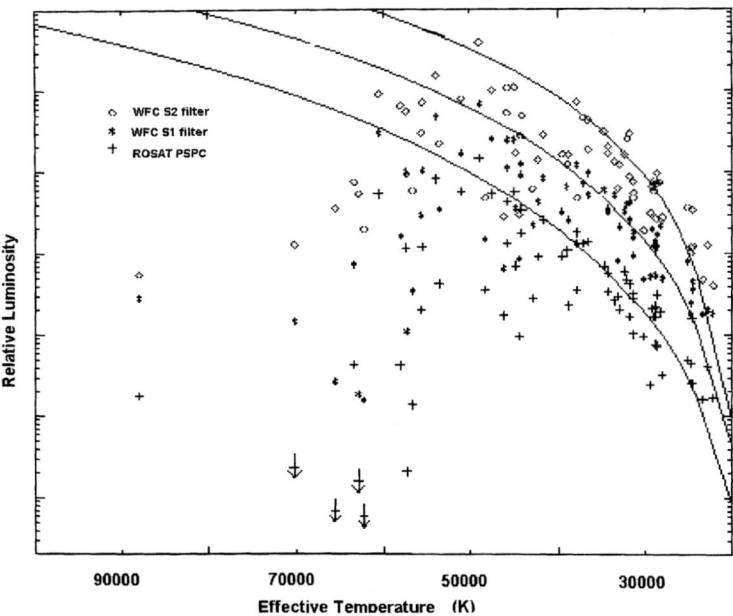

FIGURE 2. Observed X-ray and EUV intrinsic luminosity of a sample of 80 white dwarfs detected in the *ROSAT* survey, as a function of stellar temperature. The three symbols correspond to the three broad bands of the survey instruments. The solid curves correspond to the expected flux from a pure H atmosphere in each band (s2=114-180Å, s1=60-140Å, pspc=44-100Å).

such objects. In several examples, the observed emission was characteristc of that typical of the white dwarf sample. Direct confirmation of the existence of the white dwarfs is given by far-UV spectra, recorded with *IUE*, showing a characteristic steeply rising blue continuum. The probability of such a close, chance alignment of a white dwarf with a normal star is very small, indicating that these are binary systems. Several systematic searches have identified a total of more than 30 objects [5] [11] [30]. The companion stars have a wide range of spectral types, from B through to M. However, anything hotter than early A will even mask the UV white dwarf signature, requiring an EUV spectrum for a positive identification [10] [29]. This is illustrated in figure 3 which shows the composite spectrum of the A2V + DA white dwarf binary β Crt. The Lyman α line of the white dwarf, at 1216Å, is barely visible while the A companion dominates completely at the longer wavelengths.

Based on the numbers of white dwarfs in binaries detected by the *ROSAT* and *EUVE* sky surveys, compared to the isolated stars, it is estimated that some 20% of all hot white dwarfs reside in systems like these. This hidden, previously unknown population of white dwarfs is a very important one. The presence of a companion star offers the prospect of obtaining dynamical constraints of the white dwarf masses and thus testing the theoretical mass-radius relation. At the very least, the visually brighter normal star component is often listed in the *Hipparcos* catalogue, providing an accurate measurement of parallax from which the white dwarf mass and radius can be independently determined [3]. The spectral type of the companion stars also provides a limit on the nature of the white dwarf progenitors and, as a result constrain the initial final mass relation.

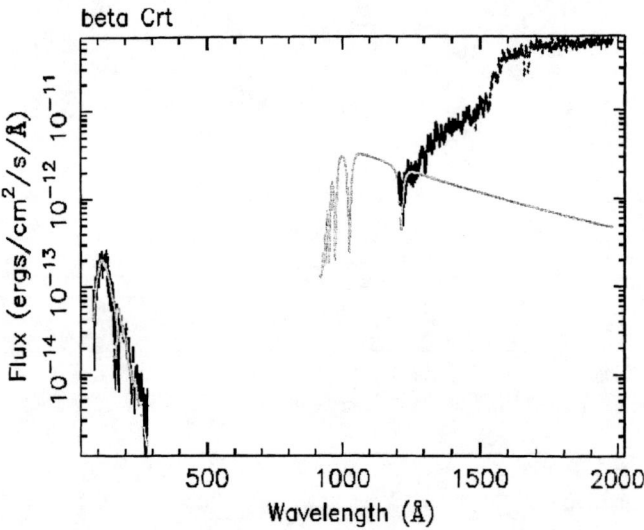

FIGURE 3. EUV and far-UV spectra of the A2V+DA binary β Crt (black error bars), compared to the prediction of a pure H model atmosphere for the white dwarf (grey curve).

SPECTROSCOPY WITH *EUVE*

The broadband photometric data obtained by *ROSAT* were able to give a global view of the incidence of opacity in white dwarf atmospheres across the whole DA population but could not yield any detail on the nature of the particular opacity sources. Spectroscopic information became available, for the brighter sources, with the launch of the Extreme Ultraviolet Explorer (*EUVE*). Covering a wavelength range from $\approx 60-750$Å, the spectrometer gives access to a region incorporating the He II Lyman series and ground state He I transitions. This provides an opportunity to probe white dwarf atmospheres for spectroscopically detectable traces of helium at levels far below those achieved by UV and optical observations. Furthermore, in atmospheres containing significant quantities of heavy element, the shape of the observed spectrum provides important constraints on the abundances.

The ionization of the local interstellar medium

The access to the He II Lyman series and ground state He I transtions provides a tremendous opportunity to probe white dwarf atmospheres for spectroscopically detectable traces of helium and examine the influence of the intervening interstellar medium. Apart from two somewhat special objects, GD50 and REJ0720−318, extensive *EUVE* observations have only obtained upper limits on the presence of photospheric helium [4] [2]. However, the spectra of approximately half a dozen stars exhibit absorption features attributable to interstellar He II. Also present is the autoionization feature of He I at 206Å. Figure 4 shows the *EUVE* spectrum of REJ2156−546 as an example.

If it is assumed that the amount of He III in the ISM is negligible, measurement of the size of the interstellar He I and He II absorption edges yields the column density of these species along the line of sight and the total column density of helium, from which the ionization fraction of helium can be determined. The decrease in the stellar flux towards the longer wavelengths is dominated by the interstellar H I opacity which can be measured by matching the observed spectrum to a combined stellar and interstellar model [2]. Making the assumption that the ratio of the total amount of hydrogen to helium has its cosmic value along the line of site allows us to predict the total H column density (10 times the total He density) and infer the amount of H II and the H ionization fraction.

This complete range of measurements is only possible for stars with a fairly narrow range of interstellar column densities, where there is sufficient material to yield measurable He II and He I edges but not so much that flux in this spectral region is completely blocked. This is illustrated in figure 4, where it can be seen that the He I 206Å autoionization feature is visible. However, the interstellar opacity at longer wavelengths renders the 504Å He I absorption edge undetectable. When the interstellar column density is much lower, $\approx 10^{18}$cm^{-2} compared to $\approx 10^{19}$cm^{-2}, longer wavelength flux can be detected and the He I edge is visible, as in HZ43 (see figure 5). Unfortunately, the He II edge then becomes too weak to detect. Although analysis of the HZ43 spectrum, the highest signal-to-noise produced by *EUVE*, hints at the presence of an edge, it is not possible to prove its existence.

In the absence of particular features, important limits can still be placed on the

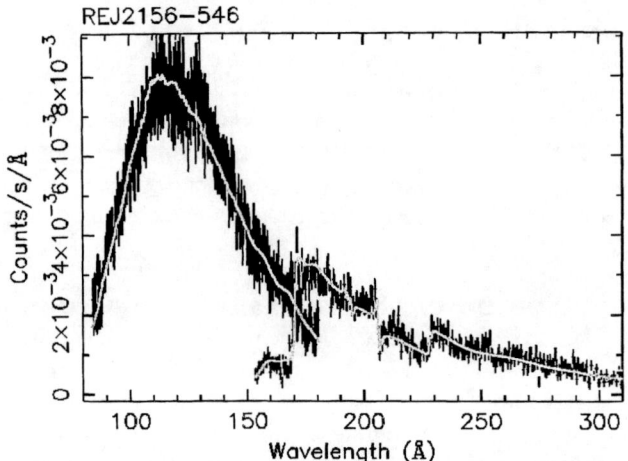

FIGURE 4. EUV spectrum of the DA white dwarf REJ2156−546 (black error bars), compared to the prediction of a pure H model atmosphere including interstellar components (grey histogram).

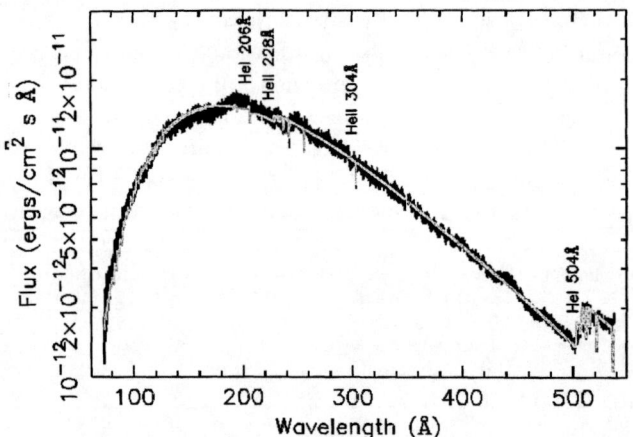

FIGURE 5. EUV spectrum of the DA white dwarf HZ43 (black error bars), compared to the prediction of a pure H model atmosphere including interstellar components (grey histogram).

column densities of the interstellar components. In a survey of 13 white dwarfs conducted by [2], both interstellar He II and He I were detected n five targets, allowing direct measurements of the He ionization fraction. Interestingly, all the measurement yielded similar values which could be treated as identical within the experimental uncertainties. Nor was there any correlation of the observed ionization fraction with direction or with the volume/column density of material along the line-of-sight. Furthermore, the limits obtained from all the other observations of lower and higher column density objects were consistent with the mean He and H ionization fractions of 0.27 ± 0.04 and 0.35 ± 0.1 respectively, calculated from the five principal measurements.

Calculations by Cheng & Bruhweiler [14] and Lyu & Bruhweiler [22] indicate that such a high level of ionization cannot be accounted for by the integrated flux of ionizing (EUV and UV) radiation from local stellar sources. Sciama [27] [28] has also proposed a more exotic source of ionizing photons originating from decaying neutrinos that might make up a dark matter component in the galaxy. An alternative mechanism for ionizing the local ISM is through shock heating. Lyu & Bruhweiler suggest that a blast wave from an ancient supernova explosion completely ionized the local interstellar cloud (LIC), in which the solar system is embedded, and that the ionization fractions we now observe are the result of subsquent recombination of the ions and electrons in the cloud. The relative H and He ionization fractions fit this picture, within the experimental uncertainties, and their absolute values indicate an elapsed time of $3 - 4$ million years since the ionizing event. The good agreement between the shock ionization model and the observations would appear to rule out the decaying neutrino idea. However, it makes the assumption that all the ionized material resides in the LIC. The *EUVE* spectra do not provide any information on the location of the absorbing material along the line-of-sight and comparison of the observed column densities with those expected from the known volume density and approximate dimensions of the LIC would tend to suggest that the ionized material may extend beyond its boundaries. The tie-in between the model and observations would then be purely coincidental. An important test will be to obtain higher resolution observations of the ionized interstellar component to acquire velocity information that can be matched to that of the LIC.

White dwarfs with heavy elements

While photospheric opacity in the form of He has only been directly detected in a handful of special cases, there are many examples of DA white dwarfs where the EUV spectrum is clearly modified by heavier elements. At the resolution available with *EUVE* nearly all the contributing lines are blended and far UV data, mainly from *IUE*, have been used to identify individual species and measure the elemental abundances. However, even with this detailed information it has proved very difficult to match the observed flux level and shape of the EUV spectra with the predictions of the theoretical stellar atmosphere calcuations.

An important prototypical heavy element-rich hot DA white dwarf is the star G191−B2B. As the visually brightest star in its class it has been the subject of most of the observational and theoretical effort. Initial studies using non-LTE models to interpret the data were not very successful [8], the synthetic spectra predicting a flux level much higher than observed at wavelengths below 250Å. These original model calculations only included those 30,000 or so Fe and Ni lines that had been observed experimentally. Lanz et al. [21] extended this work to include all the predicted transitions, amounting to some 9 million lines, yielding much better agreement with the observations. However, this apparent success hid a problem. To obtain good agreement between the models and data at the shortest wavelengths, it was necessary to incorporate an additional He II absorption component for which no direct spectroscopic evidence existed. This He II could either be interstellar/circumstellar or reside in the stellar photosphere. If interstellar, the implied He ionization fraction of $\approx 80\%$ was much larger than is typical of the local ISM. On the other hand, if the He II were photospheric, the implied abundance would produce a 1640Å line stronger than the upper limit to the line strength imposed by the *IUE* high dispersion data.

In attempt to resolve this issue, Barstow and Hubeny [6] considered a series of models where the photospheric He component was stratified. This solves the problem of the He II 1640Å line but the He II Lyman series lines were predicted to be somewhat stronger than could be comfortably accommodated by the *EUVE* spectrum. The amount of interstellar He II was also reduced to a level yielding an ionization fraction closer to that observed in other stars. While the work outlined above concentrated on matching the *EUVE* medium and long wavelength spectra, the short wavelength spectrum had not been considered at all. Yet the models predict a much higher flux in that wavelength range than observed. A complete understanding of G191−B2B requires an explanation of the short wavelength data also. A hint at a possible resolution of the problem was found by Barstow et al. [7] who noted that separate fits to the short and longer wavelength data yielded different values of the Fe abundance (figure 6), leading to the suggestion that this photospheric material might be stratified. An empirical approach was adopted, in the absence of detailed theoretical predictions of the Fe abundance at the depth of formation of the EUV continuum. The best match between the observations and data was achieved with a so-called 'slab' model comprising an outer layer of lower Fe abundance (10^{-6}) on top of a deeper region of higher abundance (4×10^{-5}, figure 7. In the best stratified models, the amount of additional He II opacity required was significantly reduced compared to the homogeneous case. It was suggested that the relative depletion of Fe in the outer layers of the envelope may be an indication of mass-loss in the form of a weak wind.

Although a two layer slab model, with a sharp division in depth between Fe abundances may seem to be a rather crude approximation, it gives significantly better agreement than models considered with a more complex structure and smoother transition. Subsequently, Dreizler and Wolff [16] have developed model atmosphere calculations which include the effects of radiative levitation self-consistently. They are able to achieve good agreement with the observations demonstrating that the depth dependence of the Fe abundance can be explained entirely as a result of the equilibrium between radiative and gravitational forces, without the need to invoke

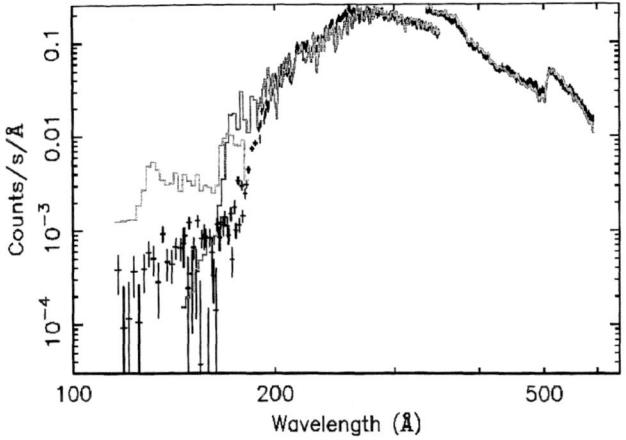

FIGURE 6. EUV spectrum of the DA white dwarf G191−B2B (black error bars), compared to the predictions of a heavy element photospheric models containing different Fe abundances (dark grey histogram - Fe/H= 3.8×10^{-6}, lighter grey histogram - Fe/H= 4×10^{-5}).

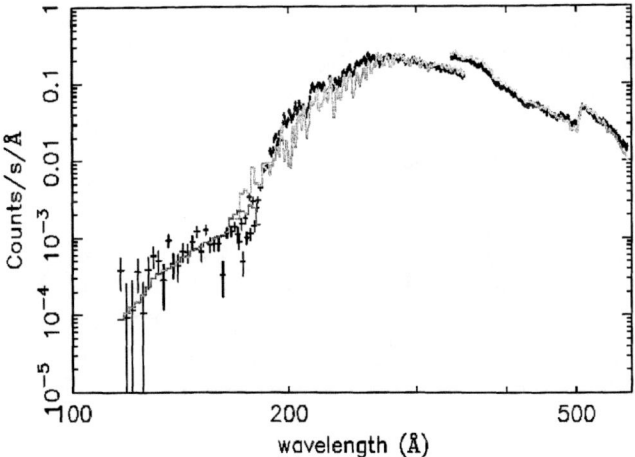

FIGURE 7. EUV spectrum of the DA white dwarf G191−B2B (black error bars), compared to the prediction of a heavy element photospheric model with stratified Fe (grey histogram).

the presence of a stellar wind. Interestingly, the predicted abundance profile of Fe is quite sharp, not too different to the much more simplistic slab model.

SUMMARY AND CONCLUSIONS

There have been dramatic advances in our understanding of the structure and evolution of hot white dwarfs made during the past 20 years. They include the discovery that heavy elements are ubiquitous in the photospheres of the hottest ($T_{eff} > 50000K$) examples and that the Fe distribution in G191−B2B, the prototype of the class, is stratified. White dwarfs have also been used as probes of the local interstellar medium, highlighting the existence of ionized helium along several lines of sight. Studies of the larger sample of white dwarfs indicate that the state of the local ISM is consistent with a more or less uniform distribution of ionization. If it is assumed that most of this material resides in the interstellar cloud surrounding the solar system, the observations are consistent with shock heating by a supernova explosion some 3-4 million years ago. Most of these important discoveries have originated from space astronomy satellites that fall into the small or medium mission class, including in particular *IUE*, the *ROSAT* WFC and *EUVE*. Follow-up work has demanded observations with larger class missions such as *HST*. However, in this field there remain important opportunities for smaller payloads. A good example is the Far Ultraviolet Spectroscopic Explorer (*FUSE*), which will provide access to the 900-1300Å window for the first time since the Copernicus mission, apart from short duration flight opportunities such as *ORFEUS*. The mass and dimensions of space-borne telescopes is typically determined by the need for a geometric collecting area as large as possible. Efficient spectrometers and detectors can mitigate this to a certain extent but in high energy telescopes, where grazing incidence optics are required, the fact that only a narrow outer annulus of the total geometric collecting area is available is leading to major large area telescopes such as *Chandra* and *XMM*. Normal incidence multi-layer telescopes offer a potential niche for smaller missions. Offering a greater geometric area than grazing incidence systems of similar aperture, they do have a disadvantage in possessing a restricted bandpass. Even so careful tuning of the wavelength ranges will potentially yield important scientific dividends. A good example is the rocket-borne J-PEX spectrometer [15], which is designed to cover the He II absorption edge at 228Å as well as several lines of the He II Lyman series, to study hot white dwarf photospheres On a longer duration mission the same waveband could also be used to study the ISM and tuning to different regions would allow studies of other sources such as stellar coronae and CVs. Since, EUV astronomy is niche science compared to mainstream X-ray astronomy it is unlikely to gain sufficient support for a major observatory class mission along the lines of *Chandra* or *XMM*. Hence, novel approaches such as that adopted by J-PEX are probably essential for future progress in this area.

REFERENCES

1. Barstow, M.A., et al., *MNRAS*, **264**, 16 (1993).
2. Barstow, M.A., Dobbie, P.D., Holberg, J.B., Hubeny, I., and Lanz, T., *MNRAS*, **286**, 58 (1997).
3. Barstow, M.A., Holberg, J.B., Cruise, A.M., and Penny, A.J., *MNRAS*, **290**, 505 (1997).
4. Barstow, M.A., Holberg, J.B., and Koester, D., *MNRAS*, **274**, L31 (1995).
5. Barstow, M.A., Holberg, J.B., Fleming, T.A., Marsh, M.C., Koester, D., and Wonnacott, D., *MNRAS*, **270**, 499 (1994).
6. Barstow, M.A., and Hubeny, I., *MNRAS*, **299**, 379 (1998)
7. Barstow, M.A., Hubeny, I., and Holberg J.B., *MNRAS*, in press (1999)
8. Barstow, M.A., Hubeny, I., Lanz, T., Holberg J.B., and Sion E.M., in *Astrophysics in the Extreme Ultraviolet*, eds. S. Bowyer and R.F. Malina, Kluwer, 203, (1996).
9. Bruhweiler, F. C., and Kondo, Y. *ApJ*, **269**, 657 (1983).
10. Burleigh, M.R., and Barstow, M.A., *MNRAS*, **295**, L15 (1998).
11. Burleigh, M.R., Barstow, M.A., and Fleming, T.A., *MNRAS*, **287**, 381 (1997).
12. Chayer, P., Fontaine G., and Wesemael F., *ApJS*, **99**, 189 (1995).
13. Chayer, P., LeBlanc, F., Fontaine, G., Wesemael, F., Michaud, G., and Vennes, S., *ApJ*, **436**, L161 (1994).
14. Cheng, K.-P., and Bruhweiler, F.C., *ApJ*, **364**, 573 (1990).
15. Cruddace, R.G., et al. , these proceedings.
16. Dreizler, S., and Wolff, B., *A& A*, in press (1999).
17. Fleming, T.A., Barstow, M.A., Sansom, A.E., Holberg, J.B., Liebert, J., and Tweedy, R., in *White Dwarfs Advances in Observation and Theory*, ed. M.A.Barstow, Kluwer, 155 (1993).
18. Holberg, J.B. et al., *ApJ*, **416**, 806 (1993).
19. Holberg, J.B., Hubeny, I., Barstow, M.A., Lanz, T., Sion, E.M., and Tweedy, R.W., *ApJ*, **425**, L105 (1994).
20. Jordan, S., Wolff, B., Koester, D., and Napiwotzki, R., *A&A*, **290**, 834 (1994).
21. Lanz, T., Barstow, M.A., Hubeny, I., Holberg, J.B., *ApJ*, **473**, 1089 (1996).
22. Lyu, C.-H., and Bruhweiler, F.C., *ApJ*, **459**, 216 (1996).
23. Marsh, M.C. et al., *MNRAS*, **287**, 705 (1997).
24. Paerels, F.B.S., Bleeker, J.A.M., Brinkman, A.C., Gronenschild, E.H.B.M., and Heise, J., *ApJ*, **308**, 190 (1986).
25. Paerels, F.B.S., Bleeker, J.A.M., Brinkman, A.C., and Heise, J., *ApJ*, **309**, L33 (1986).
26. Paerels, F.B.S. and Heise, J., *ApJ*, **339**, 1000 (1989).
27. Sciama, D.W., *ApJ*, **364**, 549 (1990)
28. Sciama D.W., *Modern Cosmology and the Dark Matter Problem*, Cambridge (1993).
29. Vennes, S., Berghoefer, T., and Christian, D., *ApJ*, **491**, L85 (1997).
30. Vennes, S., Christian, D.J., and Thorstensen, J.R., *ApJ*, **502**, 763 (1998).
31. Vennes, S., Pelletier, C., Fontaine, G., and Wesmael F., *ApJ*, **331**, 876 (1988).
32. Werner, K., and Dreizler, S., *A& A*, **286**, L31 (1994).
33. Wolff, B., Jordan, S., and Koester, D., *A& A*, **307**, 149 (1996).

Wide-field all-sky monitor for X-ray astronomy

Konstantin N. Borozdin*, William C. Priedhorsky*,
Vadim A. Arefiev**, Alexander S. Kaniovsky**, Kevin Black[†],
and Soeren Brandt[††]

NIS-2, Los Alamos National Laboratory, Los Alamos, NM 87545
**Space Research Institute, 117810 Moscow, Russia*
[†] *Goddard Space Flight Center, Greenbelt, MD 20771*
[††] *Danish Space Research Institute, DK-2100 Copenhagen, Denmark*

Abstract. The concept of using a pinhole camera as wide field-of-view detector for an X-ray all-sky monitor was first proposed by S.Holt and W.Priedhorsky in 1987 [1]. The hardware for such a monitor is ready to be launched. Here we discuss scientific tasks for such an experiment, its main parameters, and possibilities to install it on platforms/satellites of different types.

INTRODUCTION

In 1987 Holt and Priedhorsky (hereafter HP87) proposed a set of pinhole cameras as a scheme for an effective and economical X-ray all-sky monitor (ASM). Such experiment was chosen as the ASM for the international X-ray observatory Spectrum-X-Gamma, and named MOXE (Monitoring X-ray Experiment). The MOXE flight hardware has been built by the collaboration of Los Alamos National Laboratory, Goddard Space Flight Center and Moscow Space Research Institute and since 1997 has been ready for integration with the spacecraft. Six individual pinhole cameras, each with square field of view slightly larger than 90 by 90 degrees, arranged in a cubic configuration, are able to cover the whole celestial sphere.

Wide field of view of detectors allows the experiment to simultaneously cover a large fraction of the celestial sphere. This is the major difference between this instrument and earlier flown X-ray all-sky monitors, most of which were scanning devices. A scanning all-sky monitor observes only a small fraction of the sky at any particular moment. It typically observes each celestial source a few minutes per day. While this is sufficient for long-term variability studies, such an ASM is extremely ineffective for studying short-term events, e.g., X-ray bursts, fast X-ray transients, X-ray signals from gamma-ray bursts, etc. Long-term variability studies for highly variable sources (for example, black hole candidates in their low state are extremely variable on timescales of seconds) are burdened by the fact that measurements are done at random moments in time and might not represent average daily fluxes adequately. We believe that a wide-field all-sky monitor which observes a large fraction of the sky simultaneously, and hence monitors each source orders of magnitude longer than scanning narrow-field device, is able to provide better quality data for both short-term and long-term variability studies. In combination with high sensitivity, such an experiment promises to bring new important discoveries for X-ray astronomy.

SCIENTIFIC OBJECTIVES

Scientific objectives for wide-field ASM have been discussed in previous works [1–3]. It is worth while, however, to review them again in light of new information obtained in recent years in the field of X-ray astronomy.

X-ray transients

Many Galactic sources emit detectable X-ray flux only occasionally, and are called X-ray transients. They are typically binary systems composed of a degenerate object (black hole or neutron star) and a high-mass (O or B star) or low-mass (K or M dwarf) companion, and are subdivided into low-mass X-ray binaries (LMXBs)

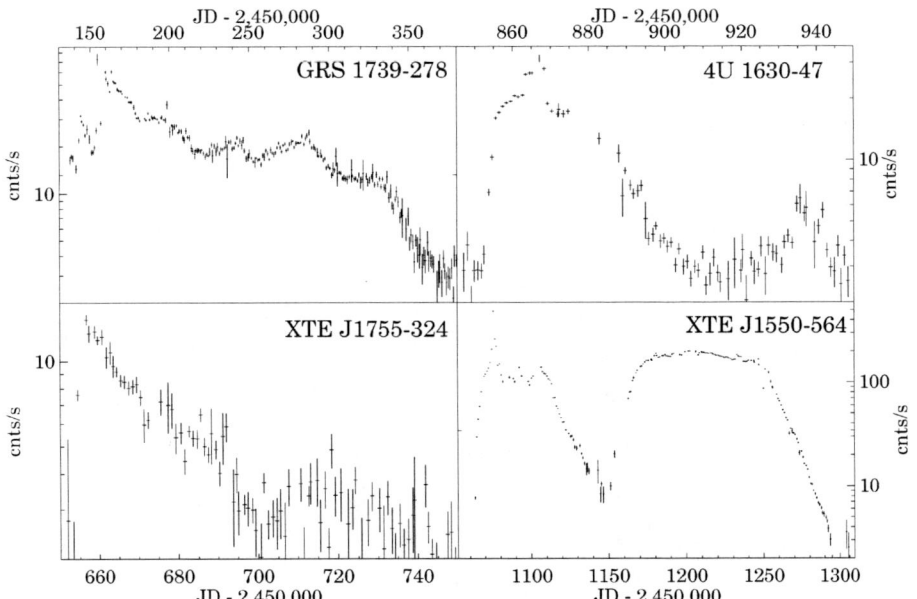

FIGURE 1. Some outbursts of X-ray Novae detected with ASM/RXTE.

and high-mass X-ray binaries (HMXBs) according to the type of the companion, and also into black hole (BH) binaries and neutron star (NS) binaries according to the type of degenerate star [4].

Most HMXB transients are Be-binary pulsars. They exhibit outbursts at regular intervals, caused by periodic encounters of a neutron star in an eccentric orbit with a stellar wind around an early-type star. The physical origin of the flux outburst, however, is not well understood, and even the equivalence of the outburst and binary cycles remains hypothetical for most objects. Most likely, the periodic outbursts result from enhanced mass transfer at periastron.

LMXB transients are characterized by episodic irregular X-ray outbursts. The recurrence time for most of these systems is estimated to be 10-100 years. The typical duration of an outburst is several weeks, but some outbursts last just few days, while some sources remain bright for years after an initial outburst. Because the time of the outburst cannot yet be predicted, an x-ray ASM is needed to trigger multiwavelength studies. Several examples of the outbursts detected with ASM/RXTE are shown in Fig. 1. LMXB transients are of special interest because most reliably identified black hole binaries belong to this class. During an outburst, the flux from the system varies by orders of magnitude, typically accompanied by dramatic changes in spectral shape. These sources are natural laboratories for studying of mass accretion onto the compact object in very different regimes.

Scientific issues addressed by the study of X-ray transients include:

- physics of the accretion onto black holes and neutron stars

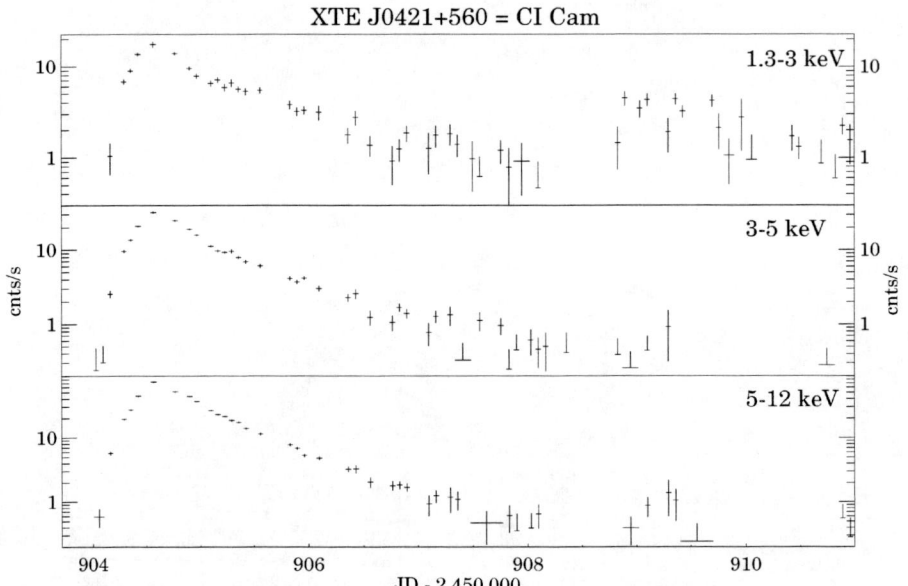

FIGURE 2. Outburst of the unusual X-ray transient XTE J0421+560(= CI Cam).

- mechanism of the outbursts
- total number of such systems in the Galaxy
- sky distribution of X-ray transients and their subclasses

Typically, 1-2 bright X-ray transients are detected per year, though it looks like there have been more detections in recent years thanks to the all-sky monitor on the RXTE satellite. The number of shorter X-ray transients (with outbursts lasting 2-3 days) could be even larger, but they have not been studied as intensively. A new generation of all-sky monitors is expected to improve significantly our knowledge of shorter and weaker transients (see Fast X-ray Transients below).

In 1998 an unusual transient XTE J0421+560 was detected [5]. The outburst was unusually short and the spectrum non-typical for earlier observed X-ray transients [6–8]. The light curve of this object in three ASM 'colors' is shown in Fig. 2. Radio observations showed evidence of an expanding quasi-spherical shell [9]. This was the first X-ray transient of this kind, and is distinctly different from earlier observed tarnsients. It was estimated that the optical counterpart of the X-ray source - CI Cam - is relatively close to the Earth (~ 1 kpc). With more sensitive all-sky monitors, such as MOXE, one may detect similar outbursts from more distant objects.

Special emphasis should be given to the earliest detection and localization of X-ray transients, in order to allow prompt follow-up observations by pointing instruments. Such observations allow better understanding of the instability leading to the outburst, and may catch the source in unusual spectral states.

The effectiveness of an all-sky monitor for the aforementioned tasks is determined by its sensitivity, simultaneous sky coverage, duty cycle, angular resolution and effective response time, which depends on the latency of data read-out and the timeliness of data analysis.

Fast X-ray transients and their relation to gamma-ray bursts

Fast X-ray Transients (FXTs) are short duration X-ray sources that have been observed by many experiments. Typical duration of the event range from seconds [10] to less than a about a day [11,12]. At present they are not associated with one particular class of emitters, but several potential types of sources are suggested to be the origin of FXTs. Their relationship to gamma-ray bursts (GRBs)is particularly intriguing. In fact FXTs resembles GRBs in many ways, save for different energy band, but FXTs are presently much less explored.

Previous surveys for fast X-ray transients, using data from HEAO-1, Ariel-V, and ROSAT, suffer from many problems. Few sources are detected, because of modest sensitivity and time-solid angle coverage. Source location errors are large, and energy and timing resolution is poor, so it has not been possible to attribute fast X-ray transients to individual counterparts, or even to identify a source population. The spatial distribution of known fast transients is too rough to distinguish between Galactic and extragalactic populations. It has been suggested that a fraction of fast X-ray transients are likely of RS CVn origin, though the rest remain unidentified. A fraction of fast X-ray transients must be counterparts of gamma-ray bursts, based on earlier results of Ginga and recent BeppoSAX observations [13–15]. Our present ignorance about fast X-ray transients leaves open the possibility of a hitherto unknown astrophysical phenomenon.

Previous estimates set the all-sky X-ray transient rate at hundreds of events per year in the 2-12 keV energy range, although only a few dozen have been detected (see Fig. 3). We intend to increase significantly the census of fast X-ray transients. We would thereby hope to better understand their statistics and distribution, distinguish Galactic and extragalactic events, study X-ray emission associated with GRBs, search for "gamma-ray quiet" GRBs, and move towards an understanding the origin of fast X-ray transients.

Some extragalactic sources like a few AGN, several sources in M31 and the SMC, Mkn421 [16,12] have shown FXT events. It was suggested that some FXT might be produced (together with a GRB) during a Type Ib/c Super Novae explosion [17]. In this case, strong beaming of up to 10-3 sr is required to allow the high energy release for high redshifted GRB's, comparable otherwise with total explosion of Super Novae. The discovering of FXT associated with such events could give direct evidence for the existence of such a process. Jet emission from quasars or AGN

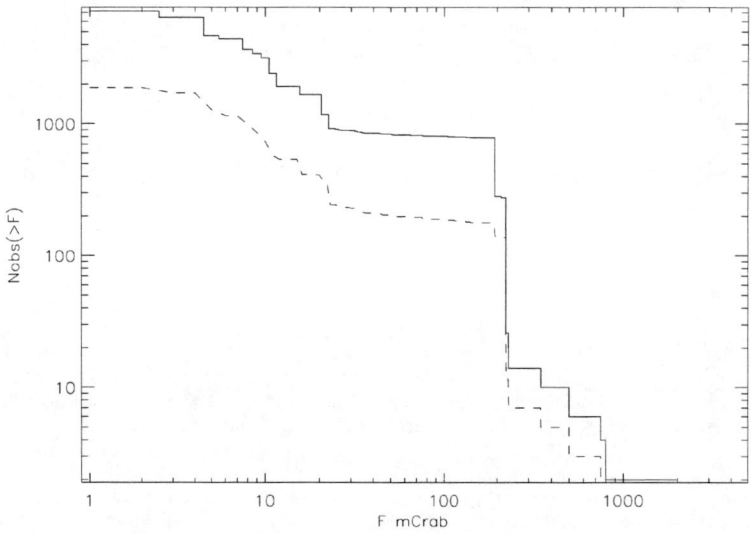

FIGURE 3. Expected number of Fast X-ray Transients (upper and lower limits) in the whole sky per year, based on data from HEAO-1, Ariel-V, Granat missions. X axis - peak flux at 2-10 keV range, Y axis - total number of X-ray transients with peak flux higher than given.

could be another source for FXTs.

The large fraction of GRBs emits soft X-ray radiation as well. The average ratio of energy emitted in the X-rays relative to the gamma rays found by Ginga [13] is 24% for a 22 GRB sample. More recently, the BeppoSAX satellite has observed several bursts in both X-rays and gamma rays [18] and discovered soft X-ray afterglow [19]. It is not yet clear whether a relativistically expanding fireball is quasi-isotropic or collimated, or if the beaming differs between the prompt emission and afterglow, although the differences for both source emission and population models are profound. It has been suggested [20] that relative beaming (prompt vs. afterglow) may be constrained by analysis of the FXT cataloged data. If a significantly larger rate of afterglow events is detected than expected from the shape of logN-logS curve for prompt GRBs of the same total fluency threshold, then afterglow beaming must be less than beaming during the burst.

On the experimental side, it remains a challenge to ensure the earliest detection of the X-ray signal from GRBs. Good sky coverage, high sensitivity and good angular resolution together with short announcement time are key elements to make all-sky monitor an effective fast X-ray transient detector. To increase the sensitivity of the instrument to FXT events one might consider replacing the pinhole with a small coded mask in the aperture.

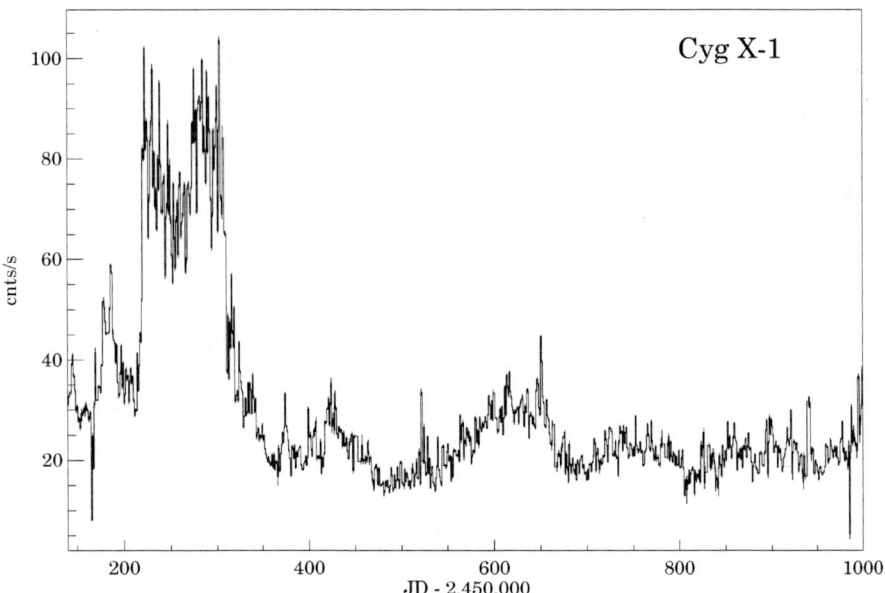

FIGURE 4. Light curve of Cyg X-1. Transitions between 'high' and 'low' states are evident.

Long-term monitoring of X-ray sources

Variability of Galactic X-ray sources provides us with valuable information regarding these sources. In fact, periodic variations, as pulsations of X-ray flux with a period of neutron star rotation, and irregular ones, such as X-ray bursts from thermonuclear flashes, are the most direct indications of the nature of the object observed. Long-term variations are harder to observe, though cycle periods up to a few years have been seen in several systems.

Cycles of 30–300 days have been confirmed for four high-mass systems, LMC X-4, Her X-1, SS433, and Cyg X-1, and are suspected in several others [21]. These cycles are observed in both the X-ray and optical bands. Some component of these systems is precessing, but we are not certain which. It could be a misaligned companion star; the outer rim of the accretion disk, driven by radiative feedback; or the compact object.

Several low-mass X-ray binaries have quasi-periodic cycles, with periods ranging from 0.5 to 2 years. The amplitude of modulation ranges between 50 and 100 percents, i.e. both "persistent" and "transient" objects fall into this class. This activity may be caused by mass transfer instabilities.

Compared to other aspects of X-ray astronomy, long-term variations have been much less intensively studied and remains poorly understood. All-sky monitors are

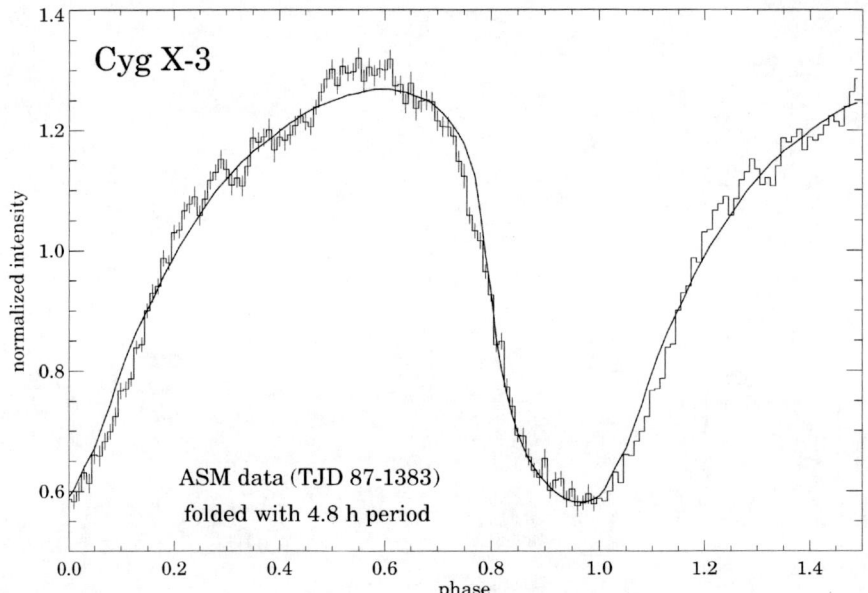

FIGURE 5. Average orbital light curve of Cyg X-3. The histogram represents ASM/RXTE data collected over period of several years and folded with Cyg X-3 orbital period. Solid line - the template for data of Copernicus satellite.

the primary source of information in this field.

State transitions

Most "persistent" Galactic X-ray sources demonstrate episodically unusual behaviour or switch from one state to another (see Fig. 4). Monitoring of these sources allows us to detect such events and find out when an observation of the source could be of special interest. In this way all-sky monitors help to define a program of observations for narrow-field instruments and enable increased scientific output of major missions, besides stimulating observations of X-ray objects at other wavelenghts.

Orbital periods

Monitoring of X-ray binaries allows us in some cases to detect their variability due to orbital motion. While orbital periods were measured for many X-ray binaries, data of all-sky monitors allows to find new, previously unknown periods. For example, periodicity of 1.4 d was reported recently in XTE J1550-564 [22], based on RXTE ASM data. ASM data allows us to study the shape of an orbital light curve and its change with time, which imposes important constraints on the parameters of the binary system, e.g., the masses of the interatiing bodies, the ellipticity of the orbit, etc. Fig. 5 illustrates an average light curve of the X-ray source Cyg X-3 built with ASM/RXTE data collected in several years. The smooth solid line shows the template, which describes data of the Copernicus satellite flown in the 1970's [23].

X-ray bursters

A wide field of view will allow us to detect numerous X-ray bursts from many bright X-ray bursters. BeppoSAX's wide field cameras, covering about 4% of the sky at any moment, were able to register more than 200 X-ray bursts from Galactic bulge sources per year [24]. With MOXE, we can measure rates of bursts from a number of bursters. New bursters may also be discovered, which will allow to identify the nature of these systems (X-ray bursts are associated with neutron stars).

Extragalactic objects

Historically all-sky monitors have studied mostly Galactic objects. Hovewer, as the sensitivity increases, more and more extragalactic objects can be observed. For the level of sensitivity which we expect to achieve, more than 50% of routinely detected objects would be of extragalactic origin. Approximately one hundred active galactic nuclei (AGN) can be monitored daily. Variability studies are essential in understanding the physics of the central regions of AGN. It is thought that the enormous energy emitted in AGN is liberated by the accretion of matter onto a massive black hole at the center of a galaxy [25]. Studies of X-ray variability provide the strongest support for this model and promise to yield fundamental new insights into the interaction of massive black holes with their environment in the cores of these galaxies [26].

PLATFORMS

Wide-field X-ray all-sky monitors, such as we are discussing here, are robust devices, without strict requirements for orientation, environment, are relatively light and have low power consumption needs, and so they are suitable for different satellite platforms. In this section we briefly describe specifics of the instrument accomodation for several platforms.

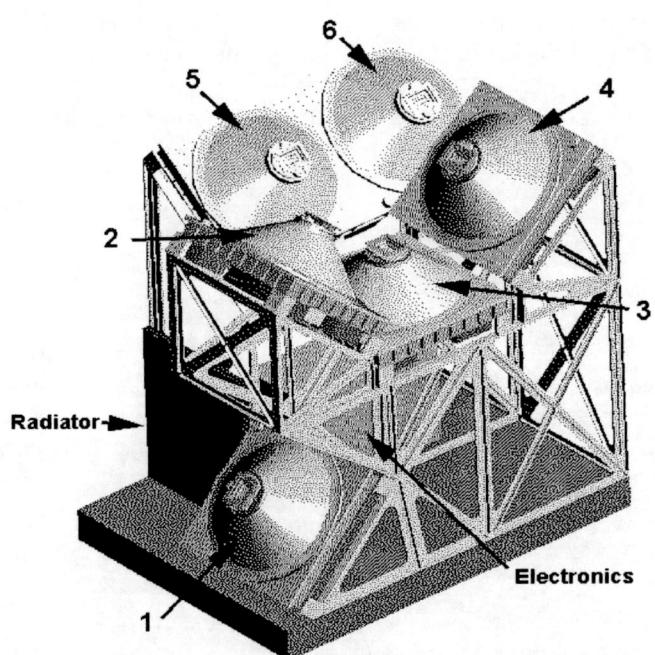

FIGURE 6. Variant of mounting for 6 wide-field cameras on ISS EXPRESS pallet.

Three-axis stabilized satellite on high-apogee orbit

Spectrum-X-Gamma under development in Russia [27] is planned to have three-axis stabilization and to be put in a high-apogee orbit, which provides low obscuration by the Earth for most of the orbit. The MOXE all-sky monitor, consisting of 6 pinhole cameras, is included as a payload for this satellite. The main advantage of this platform is an opportunity to observe almost the total sky during most of the orbit. However, it is planned that data from the satellite will be transferred on the ground only once per day, which introduces a delay for transient alerts.

International Space Station

The International Space Station (ISS) was considered as a possible platform for a wide-field ASM in the initial proposal in HP87. The experiment is small enough to fit space and weight margins for one of the places on the EXPRESS pallet of the ISS (see Fig. 6). Because of the low orbit of the space station, the instantaneous field-of-view cannot be more than a hemisphere, and further reduced by the station's modules, other experiments, and solar panels. Fields of view of modules are sufficiently overlapped, which allows increase in sensitivity and source localization accuracy (Fig. 7). Real-time telemetry link for most of the time can make it possible to detect bright transient events in seconds. Rotation of the space station allows us to observe the whole sky during one revolution of the station (1.5

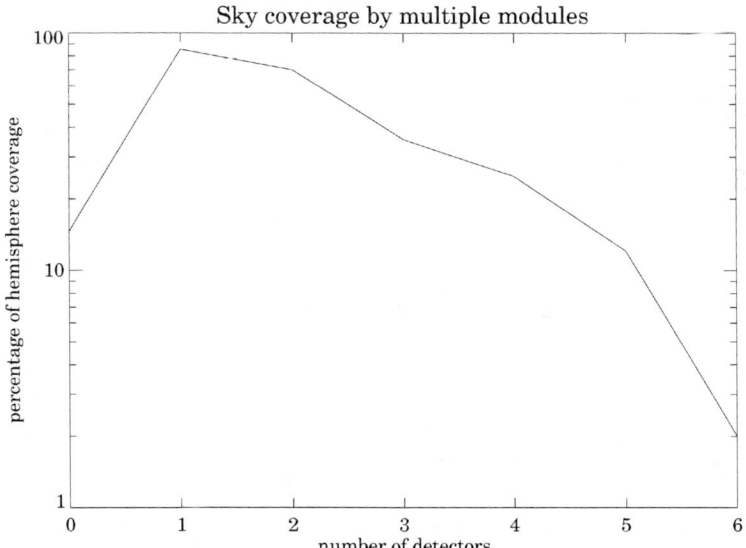

FIGURE 7. Sumaltaneous coverage of sky by 6 wide-field detectors installed as shown in Fig.6.

TABLE 1. Parameters of modern X-ray monitors

	ASM/RXTE	MOXE/SRG	MAXI/JEM-X	MOXE/ISS
Instant sky coverage, ster	0.5	12.	0.1	5.4
Duty cycle[a]	0.05	~1.	0.006	~0.3
Sensitivity, mCrab/1.5h	30	7	10	10
Localization accuracy	3'x15'	30'x100'	20'x50'	30'x30'
Energy band	1.3-12 keV	2-20 keV	0.5-30 keV	2-20 keV
X-ray burst detection probability[b]		0.8	0.009	0.3
X-ray Nova detection probability[c]		~1	~1	~1
Typical alarm time, h		30	24	~1[d]
Launch date[e]	12/30/1995	2001	2004	-
Reference	[28]	[2,3]	[29]	-

[a] time of point source observation divided by time needed to cover whole sky
[b] 1 Crab source with 10 sec duration
[c] 100 mCrab source with 1 day duration
[d] can be order of seconds, if onboard processing implemented
[e] actual date for ASM/RXTE, expected in other cases

hours) and achieve more uniformal measurements for each source independent of its position on the detector at any particular moment.

Small mission

A wide-field ASM can be also a payload for a small autonomous satellite. Contrary to the space station, it would be easy to avoid any additional obscurations in this case, and increase field of view to the maximum possible. Orbit of the satellite and opportunities of real-time connection to the ground would be important criteria for such a mission. In the ideal case, capabilities of the experiment could be superior in comparison to other platforms.

PARAMETERS OF THE EXPERIMENT

Wide-field of view makes the all-sky monitor which we discuss here substantially different from any scanning instruments. This allows it to detect much more fast X-ray transients, X-ray bursts and follow changes in bright X-ray sources with finer time scale. The sensitivity of the experiment is higher than any ASM launched during the past 40 years of X-ray astronomy. Actual capabilities of the experiment depend on the platform where it will be installed.

The comparison of estimated parameters of two versions of wide-field monitor - MOXE on SRG and on ISS - with other X-ray ASMs is presented in Table 1.

CONCLUSION

A pinhole camera is an elegant but effective approach to an x-ray all-sky monitor(ASM) [1]. Six pinhole cameras, each with 90x90 degrees field of view, are able to monitor the whole sky simultaneously. The instrument does not require pointing, is robust against contamination, and requires modest resources (120 kg, 50 Watts, 10 kpbs). Nonetheless, it would be more sensitive than any previous all-sky x-ray monitor. By continuously monitoring the entire unocculted sky, this instrument would be sensitive to changes at all timescales. Besides monitoring the brightest few hundred x-ray sources, including about a dozen active galactic nuclei, it would be uniquely sensitive to fast x-ray transients, unlike any scanning instrument. We would expect to detect several hundred events per year with timescales from a minute to a day, and better understand their correlation with magnetic activity on nearby stars and with gamma-ray bursts. We expect to detect about 50 gamma-ray bursts per year and locate them to 1 square degree, This would be the first x-ray selected survey of gamma-ray bursts. We discussed possible implementations of the instrument for different satellite platforms.

ACKNOWLEDGEMENTS

For this work we used results provided by the ASM/RXTE teams at MIT, and at the RXTE SOF and GOF at NASA's GSFC.

REFERENCES

1. Holt S.S., and Priedhorsky W.C., *Space Sci. Rev.* **45**, 269 (1987).
2. in't Zand J.J., Priedhorsky W.C., Moss C.E., Fenimore E.E., Black J.K., Kelley R.L., Stilwell D.E., Birsa F.B., Borozdin K.N., and Arefiev V.A., *Proc. SPIE*, **2279**, 458 (1994).
3. Brandt S., Priedhorsky W.C., Fenimore E.E., Moss C.E., Black J.K., Kelley R.L., Stilwell D.E., Birsa F.B., Borozdin K.N., Kaniovsky A.S., Arefiev V.A., and Efremov V.V., *Physica Scripta*, **T77**, 21 (1998).
4. Tanaka Y., and Shibazaki N., *Annu. Rev. Astron. Astrophys.*, **34**, 607 (1996).
5. Smith D., Remillard R., Swank J., Takeshima T., and Smith E., *IAU Circ.*, **6855**, 1 (1998).
6. Orr A., Parmar A.N., Orlandini M., Frontera F., Dal Flume D., Segreto A., Santangelo A., and Tavani M., *Astron. Astrophys.*, **340**, L19 (1998).
7. Ueda Y., Ishida M., Inoue H., Dotani T., Greiner J., and Lewin W.H.G., *Ap. J.*, **508**, L167 (1998).
8. Revnivtsev M.G., Emel'yanov A.N., and Borozdin K.N., *Astron. Lett.*, **25**, 294 (1999).
9. Mioduszewski A., *LAAstro Seminar*, April 28 (1999).
10. Belian R.D., Conner J.P., and Evans W.D., *Ap. J.*, **207**, L33 (1976).

11. Castro-Tirado A.J., *Ph.D. Thesis*, Univ. of Copenhagen (1994).
12. Pye J.P., and McHardy I.M., *M.N.R.A.S.*, **205**, 875 (1983).
13. Strohmayer T.E., Fenimore E.E., Murakami T., and Yoshida A., *Ap. J.*, **500**, 873 (1998).
14. in't Zand J.J.M., Amati L., Antonelli L.A. et al., *Ap. J.*, **505**, L119 (1998).
15. in't Zand J.J.M., Heise J., van Paradijs J., Fenimore E.E., *Ap. J.*, **516**, L57 (1999).
16. White N.E., Angelini L., and Giommi P., *All-Sky X-ray Observations in the Next Decade*, eds. M.Matsuoka, and N.Kawai, RIKEN, Japan, 41 (1997).
17. Wang L., and Wheeler J.C., *Ap. J.*, **504**, L87 (1998).
18. Piro L., Heise J., Jager R. et al., *Astron. Astrophys.*, **329**, 906 (1998).
19. Costa E., Frontera F., Heise J. et al., *Nature*, **387**, 783 (1997).
20. Grindlay J.E., *Astrophys. J.*, **510**, 710 (1999).
21. Priedhorsky W.C., and Holt S.S., *Space Sci. Rev.*, **45**, 291 (1987).
22. Cui W., Wen L., Zhang S.N., Wu X.-B., *IAU Circ.*, **7191**, 1 (1999).
23. van der Klis M., and Bonnet-Bidaud J.M., *Astron. Astrophys.*, **214**, 203 (1989).
24. in't Zand J.J.M., *Private communication*, (1999).
25. Rees M.J., *Annu. Rev. Astron. Astrophys.*, **22**, 471 (1984).
26. Mushotzky R.F., Done C., and Pounds K.A., *Annu. Rev. Astron. Astrophys.*, **31**, 717 (1993).
27. Sunyaev R.A., Borozdin K.N., Lapshov I.Yu., and Terekhov O.V., *Preprint IKI AN USSR*, **Pr-1632**, 1 (1990).
28. Levine A.M., Bradt H., Cui W., Jernigan J.G., Morgan E.H., Remillard R., Shirey R.E., and Smith D.A., *Ap. J.*, **469**, L33 (1996).
29. Matsuoka M., Kawai N., Mihara T. et al., *Proc. SPIE*, **3114**, 414 (1997).

BALLERINA - Pirouettes in Search of Gamma Burst Sources

Søren Brandt and Niels Lund[1]

Danish Space Research Institute
Juliane Maries Vej 30, DK-2100 Copenhagen Ø, Denmark

Abstract. The cosmological origin of gamma-ray bursts (GRBs) has now been established with reasonable certainty. Many more bursts will need to be studied to establish the typical distance scale, and to map out the large variability in properties, which have been indicated by the first handful of events. We are proposing BALLERINA, a small satellite to provide accurate gamma burst positions at a rate an order of magnitude larger than from Beppo-SAX. On the experimental side, it remains a challenge to ensure the earliest detection of the X-ray afterglow. The mission proposed here allows for the first time systematic studies of the soft X-ray emission in the time interval from only a few minutes after the onset of the burst to a few hours later. In addition to positions of GRBs with accuracy better than 1' reported to the ground within a few minutes of the burst, essential for follow-up work, BALLERINA will on its own provide observations in an uncharted region of parameter space. Secondary objectives of the BALLERINA mission includes observations of the earliest phases of the outbursts of X-ray novae and other X-ray transients. BALLERINA is one of four missions currently under study for the Danish Small Satellite Programme. The selection will be announced in 1999 for a planned launch in 2002-2003.

INTRODUCTION

In recent years the concept of small satellites has received considerable attention as a way to increase the relative return of investments in space research. The idea of *smaller and better* is sometimes taken as a dogmatic truth. However, in many cases this approach is not viable. The purpose of this workshop is to identify scientific topics, which are well served by small missions. The feasibility of addressing a certain set of scientific questions through a small mission depends on many parameters. One such parameter is the lead time. In scientific fields undergoing rapid change it is of particular importance to shorten the time between mission conception and actual implementation.

A scientific field, which has been revolutionized over the past few years, craving answers to unforeseen questions, is the mystery of the origin of cosmic gamma-ray

---

[1] on behalf of the BALLERINA consortium

bursts. However, it is not the purpose of this paper to analyze the topic of small missions in general, but to present BALLERINA as a mission, which has a science case very well suited for a small mission.

THE BALLERINA MISSION

BALLERINA has been proposed [1] for the Danish Small Satellite Programme [2], and is one of four missions currently under study. The selection will be announced in 1999 for a planned launch in 2002-2003. The development of BALLERINA will depend on a significant heritage from the ØRSTED mission, the first Danish satellite [3].

The BALLERINA Payload Concept

The BALLERINA payload consists of two main instruments: the all–sky monitor and the narrow field telescope-assembly consisting of a grazing incidence X–ray telescope and a collimator telescope feeding a common focal plane detector.

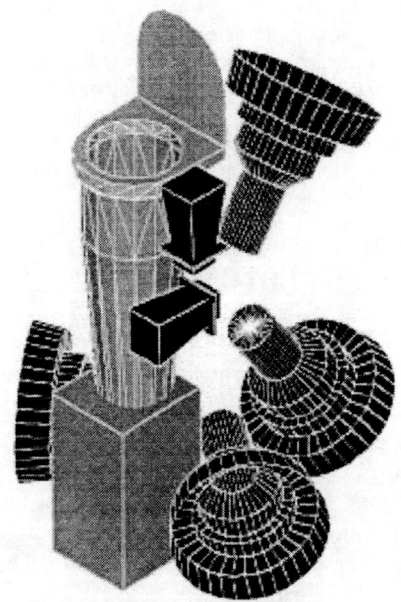

FIGURE 1. The BALLERINA payload concept. The four identical WATCH wide field X-ray cameras cover the full sky. A grazing incidence X-ray telescope and a collimator telescope feed a common focal plane detector. In addition, two perpendicular star trackers are shown.

TABLE 1. Summary of the BALLERINA payload capabilities. Note that the number of bursts detected with the pointed instruments does not include bursts detected by other missions, and subsequently studied with BALLERINA.

	Energy Range	Effevtive Area	Field of View	Localisation Accuracy	Bursts per Year
All–sky Monitor	6–120 keV	45 cm^2	4π	1°	≈ 80
Telescope	0.5–2.0 keV	50 cm^2	2°	0.5′	≈ 70
Collimator	0.2–15 keV	20 cm^2	2°	–	≈ 70

The times and positions of gamma-ray bursts are random. The idea is to detect the gamma-ray burst with the all-sky monitor, and then rapidly re-orient the spacecraft to observe the burst position with the more sensitive and accurate narrow field X-ray telescope. The satellite operations are autonomous, but the positions of bursts are distributed to the ground in real time for rapid follow-up. An outline of the payload accommodation is shown in Fig. 1. The capabilities of the BALLERINA payload is summarized in Table 1

The all-sky monitor consists of four rotation modulation collimator (RMC) instruments, WATCH, each covering slightly more than one quarter of the sky. The performance of WATCH as a gamma-ray burst detector is well established from two previous satellite missions [4–7], and we expect the all-sky monitor to detect ≈ 80 bursts per year.

WATCH can localize a burst to better than 1° (2σ). The burst position can be calculated on-board in less than one second after collection of the burst data. This coarse, initial position will immediately be transmitted to the ground.

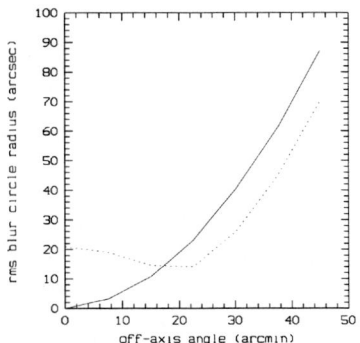
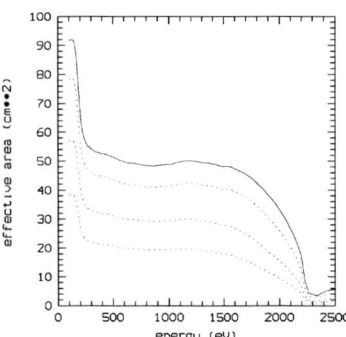

FIGURE 2. Expected performance of the BALLERINA X-ray telescope. To the left is shown the rms blur circle radius as function of off-axix angle. The solid curve shows the nominal focus, and the dotted line shows a shifted focal plane optimized for off-axis resolution. To the right is shown the effective area as function of energy. The solid line is for on-axis X-rays, and the dotted lines are for 15′, 30′, and 45′off-axis, respectively.

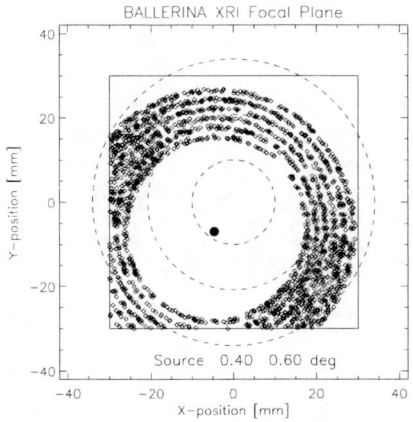

FIGURE 3. Simulation of the response of the BALLERINA X-ray collimator to a source ≈ 0.7° off-axis. The box indicates the boundary of the 6×6 cm^2 CCD. The inner circle indicates the area reserved for the focussing telescope, thus never illuminates through the collimator. The black dot shows the position of the image of the source through the telescope.

Based on the initial position, the spacecraft will swiftly reorient itself to point the X-ray telescope to the source region. The slew rate is ≈ 4°/s or 90° in about 20 seconds. A star tracker will determine the satellite attitude to better than 20″. It is expected to react on all burst triggers with an angle of at least 45° away from the Sun. It is expected to follow-up on ≈ 70 bursts per year with the pointed telescope.

The telescope-assembly combines two instruments using a common large format CCD detector. The first instrument is a grazing incidence telescope having a field of view of 2° and a 60 cm focal length. The energy passband of the telescope is 0.2–2 keV, with an effective area on axis of ≈ 50 cm^2 (see Fig. 2). The telescope will localize the gamma-ray burst source to be better than 0.5′. The X-ray telescope design is based on heritage from the ABRIXAS mission [8].

The second instrument is a simple collimator utilizing the outer part of the CCD-detector, not being used by the X-ray telescope, to view a 2° region containing the gamma-burst source. This collimator instrument will exploit the full 0.2–15 keV energy range of the CCD, and provide time-resolved spectra of the afterglow during its early, intense phase. An example of the response of the collimator is shown in Fig.3.

The BALLERINA Platform

The BALLERINA spacecraft will be three axis stabilized, with a pointing accuracy of 20″. A unique feature is the use of the RMC of the 4 wide field cameras

TABLE 2. Summary of the BALLERINA spacecraft resources.

Instrument	Power (Watt)	Data Rate (kbit/s)	Weight (kg)
All–sky Monitor	18	1.2	16
X-ray Telescope and CCD	15	0.5	27
2 Star trackers	7	0.3	4
Platform and Structure	22	0.2	46.5
Total	62	2.2	93.5

as momentum wheels. The feasibility of this dual use has been studied based on the performance of the WATCH monitor on the EURECA mission. The size of the satellite has been limited to $60 \times 60 \times 80 cm^3$ and a weight below 100 kg to enable a piggy-back launch opportunity, and to reduce the moment of inertia for rapid re-orientations. The baseline for the re-orientation capability is a slew-rate of $\approx 4°/s$ with a goal of achieving $\approx 8°/s$. Table 2 provides a summary of the BALLERINA resources.

The preferred orbit for BALLERINA is the high-inclination Molniya orbit. This highly elliptical orbit has a 63° inclination with a period of 12 hours. This orbit is chosen because it provides a reasonable operational efficiency with an estimated duty cycle of $\approx 60\%$, and continuous telemetry visibility from Danish ground stations during the useful part of the orbit. Most importantly, affordable launch opportunities have been identified for secondary payloads on the Russian Soyuz/Fregat launcher.

STUDYING GAMMA BURSTS WITH BALLERINA

Gamma-ray bursts (GRBs) have ranked among the greatest mysteries of astrophysics since their discovery three decades ago [10]. Developments during the past two years with the discovery of X-ray [11] and optical [12] afterglows have further ignited the imagination of scientists, as well as the general public. Observational insight into the phenomenon has accelerated quickly, and theoretical interpretations are following suit. Much is indeed already known about the physical conditions at late epochs (hours to days after the events), but the interplay between observation and theoretical modeling is not very advanced with regards to the very early phases. In particular, very little is known about the workings of the inner engine powering the gamma-ray burst. BALLERINA will contribute to all aspects of these investigations. The BALLERINA mission will study the early evolution of the X-ray afterglow to exploit this recently discovered phenomenon as a channel of information about the nature and origin of gamma-ray bursts.

BALLERINA will determine, in real time, the positions of cosmic gamma-ray bursts to a precision better than $0.5'$, and transmit these positions to the ground, thus allowing immediate follow-up in the optical, infrared, and radio wavebands.

It is not easy to predict how gamma-ray burst astrophysics will develop during the years up to the launch of BALLERINA. The current pace of discoveries may not be sustained, but there is no indication, and no expectation, that we should be anywhere close to 'solving' the enigma.

Gamma-ray bursts

The "standard model" of a GRB is a catastrophic event involving compact objects - neutron stars or black holes. In a fraction of a second, the source must liberate the equivalent of about one solar mass as electromagnetic radiation. The merging of the components of a binary neutron star system is one possibility, and the steady loss of orbital energy through gravitational radiation will eventually lead to merging of the components. The process may appear deceptively simple, but the highly complex time histories of GRBs are not easily explained.

GRBs are now known to occur at cosmological distances with typical redshifts in the range z $\approx$ 0.5 − 4 or more. These large distances imply that huge amounts of energy are released. The recent event, GRB 990123 has been estimated, under the assumption of isotropic emission, to release about 4×10^{54} erg [9], corresponding to 1000 times the energy radiated as photons in either supernova Type I or II.

There are two broad families of models for GRBs related to the end stages of massive stars; both predict that GRBs should trace the star-formation rate in the Universe (e.g. [13]), and both give specific, testable predictions about the locations of the GRBs relative to their host galaxies. The first is the exploding object family, which includes the "failed supernova" model of Woosley [14] and the "hypernova" model of Paczynski [15]. These models predict that the progenitors of GRBs are short-lived objects with low space velocities, and that GRBs should be found within $\approx$0.5 kpc of the star-forming regions, where the progenitors were formed. The second family of models is that of binary compact objects, which includes the binary neutron star (NS-NS) model of Narayan et al. [16] and the merger of a black hole and a neutron star (BH-NS) model of Paczynski [17]. These models predict that GRB progenitors will have high space velocities, due to two supernova explosions having occurred in the progenitors' binary system, so the GRB could be found at a significant distance from the star-forming region, where the progenitors were formed.

Whatever the initial mechanism behind the burst, the end result is believed to be a fireball expanding at relativistic speeds into the surrounding interstellar medium. In a shock front, that forms when the fireball sweeps up the medium, and electrons are continuously accelerated to high energies. In the standard model a magnetic field causes the electrons to produce synchrotron radiation. The instantaneous spectrum is determined by the physical conditions at the shock front, while the light curve of the afterglow and the spectral evolution is determined by the hydrodynamic evolution of the fireball. The fireball model has successfully been used to account for general properties of the afterglow following the burst.

If the expanding fireball is confined to a cone in the form of a jet pointing at us, the energy requirements are considerably relaxed. Given that the expansion is highly relativistic, it is not possible to distinguish between a spherical fireball and a jet, if the relativistic beaming angle is smaller than the opening angle of the jet. However, as the expanding fireball sweeps up material from the surroundings and decelerates, an observable break in the light curve should develop, when the relativistic beaming angle grows larger than the jet opening angle. This may have been observed in GRB 980519 [18] and is the most straightforward explanation for the break in the light curve of GRB 990123 [19].

BALLERINA: Science with Bright Bursts

During its lifetime, BALLERINA will detect, localize and image a considerable number of bright gamma-ray bursts. We expect to detect and study in detail ≈ 70 GRBs per year. This will over the two–year nominal lifetime of BALLERINA increase the number of well-studied bright events by at an order of magnitude, from about ten to more than a hundred. The rapid reorientation of BALLERINA will allow detailed investigations of the time resolved spectral evolution of GRBs. The brightness of the BALLERINA bursts, combined with the fast notification to the ground, ensures detailed X-ray studies with the on-board telescope and collimator, as well as successful, rapid ground-based optical/IR follow-up observations. BALLERINA will observe many more bright GRBs than any other current, approved, or proposed mission.

In the current sample of 9 GRBs with known redshifts, there is, as yet, no indication of a correlation between brightness and distance. In other words, because of the enormous dispersion in the intrinsic energy output, or because of the anisotropic emission pattern, our sample of brighter bursts will still include events at high redshifts.

The aim of the BALLERINA mission is to advance our understanding of the following key issues significantly:

- The origin of GRBs and the nature of their progenitors

- The relation between the GRB and the X-ray afterglow

- The physics of the afterglow

- The geometry of the gamma-ray emission and the afterglow radiation

- The birth places of GRBs - their environments and hosts

- The use of GRBs as cosmological probes

- The redshift and luminosity distribution of GRBs

- The star-formation history of the Universe

We will address these scientific problems following two broad avenues, which define the scope of the BALLERINA mission: the spectral and temporal study of the X-ray afterglow, and its use for localization of GRBs for ground-based follow-up.

ADDITIONAL SCIENTIFIC OBJECTIVES

Transient Alerts and Follow-up Observations

BALLERINA will act as an all-sky monitor for bright X-ray transients, and will be able to do follow-up observations in the soft X-ray band, and to provide alerts for multi-wavelength campaigns. This will be particularly significant for the INTEGRAL mission, where some of the most exciting science may come from observations of the few bright X-ray novae to appear during the INTEGRAL lifetime. But other missions such as HST, XMM and AXAF, Astro-E, SXG, and ground-based observers will benefit as well.

BALLERINA will also be unique as a rapid response telescope for targets of opportunity provided by other observers.

X-ray Novae

X-ray novae were discovered more than twenty years ago [20]. These double-star systems are hosts to the most promising black hole candidates in our Galaxy, and have, as such, attracted considerable attention, as they provide excellent laboratories for the study of black holes. In order to better understand the instability leading to these outbursts, more detailed observations during the earliest phases of the outbursts are required [25]. BALLERINA will provide precise positions immediately after the beginning of the outburst, and will be able to track the early evolution simultaneously in soft and hard X-rays. We expect on average to discover 1-2 bright X-ray novae per year.

We may exemplify the type of event, which can be discovered and monitored by BALLERINA, by the transient GRS 1915+105. This source was discovered by WATCH on the GRANAT mission in 1992 [21]. This black hole candidate has been one of the most studied X-ray sources of the decade, as it exhibits a multitude of extreme behaviors. For example, it was the first Galactic source to show super-luminal mass ejections [22]. Its highly complex X-ray lightcurve provides an excellent laboratory for putting any theory of accretion to the test [23]. We may speculate, that the understanding of the physics of the jets of this micro-quasar holds some clues to the understanding of the inner engine of GRBs.

Fast Transients

With continuous coverage of the sky (as opposed to all-sky monitors of the scanning type) the WATCH all-sky monitor will enable detection and follow-up of some so called "fast transients"; sources in our own Galaxy which flares for a few minutes or hours [24]. Little is known about these sources, since they have previously only been detected serendipitously, without possibility for systematic follow-up observations. The brighter of these events can easily be recognized by the BALLERINA on-board software, and the satellite will autonomously issue alerts to the ground and initiate observations with the X-ray telescope, in just the same way as in the case of a gamma-ray burst. A handful of transients of duration of less than a day were detected with WATCH on GRANAT, but the sources were not uniquely identified due to the lack of follow-up observations. BALLERINA will be ideally suited to the observation of this type of sources.

Persistent X-ray Source Monitoring

Between the GRB observations, the X-ray telescope may be used for monitoring programs of AGNs and galactic X-ray sources. Much information about X-ray binaries is derived from the study of their time variability. We will monitor the periods of X-ray pulsars; study the X-ray flux and spectrum as function of orbital phase; observe the super-orbital variability seen in some sources, yet to be discovered in others.

Solar Flares

The all-sky monitor on BALLERINA will continuously monitor the sun for the occurrence of solar flares. Due to their high brightness, the flares can be located on the solar disk to a precision of better than 10' based on the WATCH data alone. The flare data will also be subjected to a statistical study to investigate the frequency distributions of solar X-ray parameters. This will continue and extend similar work based on the previous WATCH solar flare catalogue [26,27].

COMPLEMENTARY MISSIONS

There is currently no other approved mission, which will address the issue of the spectral and time evolution of the X-ray afterglow following a gamma-ray burst.

The ongoing mission, Beppo-SAX, and two missions in preparation, HETE-II (launch 2000) [28] and INTEGRAL (launch 2001) [29], will provide good positions and allow efficient optical follow-up. These missions have smaller fields of view and will therefore provide a lower rate of localizations than BALLERINA – in particular if we consider only bright bursts. Neither HETE-II nor INTEGRAL will carry a focusing X-ray telescope, and hence they will be limited in their ability to contribute to the study of the X-ray afterglow.

The Swift mission [30] being proposed for the next NASA MIDEX selection is addressing similar questions as BALLERINA, and will be using a similar concept. However, there are appreciable differences in size and approach of the two missions. Stated briefly the Swift mission will carry a large area monitor with a field of view of 2 steradians (1/6 of the BALLERINA monitor system). Swift will also carry an X-ray telescope, and an optical monitor. Swift will perform rapid re-orientations to carry out fast follow-up observations very similarly to BALLERINA. The reorientation times are similar for the two missions.

We may compare the science of BALLERINA and Swift. BALLERINA will study about 70 bursts/year from the bright end of the luminosity distribution. Swift will go much deeper than BALLERINA, and is expected to detect 300 bursts per year from a limited region of the sky. The luminosities of the Swift bursts are expected to extend well below the faint end of the currently known burst size distribution. However, the brighter bursts may be more interesting, as they will allow more detailed follow-up observations. Regarding fainter bursts, BALLERINA can expect to receive an additional 20–30 burst triggers per year from INTEGRAL. These triggers may extend to even fainter bursts than detectable by Swift.

If the operational lifetime of BALLERINA and Swift will overlap, there will be only a small overlap (16% of the BALLERINA bursts) between the burst samples observed by BALLERINA and by Swift, as illustrated in Fig.4.

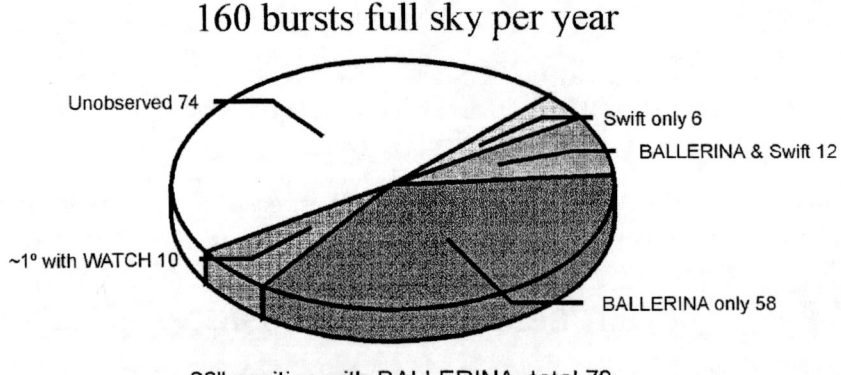

FIGURE 4. Comparison of the BALLERINA and Swift capabilities in observing the brighter GRBs. If we consider the sample of the ≈ 160 brightest bursts per year, BALLERINA will be able to do detailed observations of ≈ 70 bursts. Swift will be able to observe ≈ 18 burst from this sample.

CONCLUSION

The rapidly changing field of gamma-ray burst research is an excellent example, where the concept of a small mission promises a high scientific return. We have presented the BALLERINA mission, combining payload and platform elements with a proven track record into a powerful tool to study the early phase of GRB afterglows.

REFERENCES

1. Lund, N., and the BALLERINA Consortium, Danish Space Research Institute (1998) http://www.dsri.dk/dssp/Ballerina_Proposal_20.pdf
2. http://www.dsri.dk/dssp/DSSP-homepage.uk.html
3. http://web.dmi.dk/fsweb/Projects/oersted/homepage.html
4. Lund, N., *SPIE Proc.*, **597**, 165 (1986)
5. Brandt, S., PhD Thesis, Danish Space Research Institute (1994)
6. Castro-Tirado, A.J., PhD Thesis, Danish Space Research Institute (1994)
7. Sazonov, Y. et al., *Astronomy & Astrophysics Supplement* **129**, 1 (1998)
8. Egle, W.J., et al., *Proc. SPIE*, **3444**, 359, (1998)
9. Andersen, M.I., et al. *Science*, **283**, 2075 (1999)
10. Klebesadel, R.W., et al., *Astrophysical Journal*, **182**, L85 (1973)
11. Costa, E., *Nature*, **387**, 783 (1997)
12. van Paradijs, J., et al., *Nature*, **386**, 686 (1997)
13. Wijers, R. A. M. J. et al., *MNRAS*, **294**, 113 (1998)
14. Woosley, S.E., *Astrophysical Journal*, **405**, 273 (1993)
15. Paczynski, B., *Astrophysical Journal*, **494**, L45 (1998)
16. Narayan, R., et al., *Astrophysical Journal*, **395** L83 (1992)
17. Paczynski, B., *Acta Astronomica*, **41**, 257 (1991)
18. Sari, R., Piran, T., and Halpern, J., *Astrophysical Journal*, **519**, L17 (1999)
19. Castro-Tirado, A.J., et al., *Science*, **283**, 2069 (1999)
20. Elvis, M. et al., *Nature*, **257**, 656 (1975)
21. Castro Tirado, A.J., et al., *Astrophysical Journal Supplement* **92**, 469 (1993)
22. Mirabel, I.F, and Rodriguez, L.F., *Nature*, **371**, 46 (1994)
23. Yadav, J.S., et al., *Astrophysical Journal*, **517**, 935 (1999)
24. Pye, J.P., and McHardy, I.M., *MNRAS*, **205**, 875, (1983)
25. Lund, N., *Astronomy & Astrophysics Supplement* **97**, 289 (1993)
26. Crosby, N. et al., *Astronomy & Astrophysics Supplement*, **97**, 309 (1998)
27. Crosby, N. et al., *Astronomy & Astrophysics*, **334**, 299 (1998)
28. http://space.mit.edu/HETE/
29. http://astro.estec.esa.nl/SA-general/Projects/Integral/integral.html
30. http://swift.gsfc.nasa.gov/

High Resolution X-ray Imaging

Webster Cash

University of Colorado, Boulder, CO 80309

Abstract. Large collecting area is the domain of large missions, but high resolution x-ray images can be captured with small instruments in small missions. We discuss techniques for achieving high resolution x-ray images including high resolution telescopes and interferometry. We discuss the likely limitations within the confines of small spacecraft.

HIGH RESOLUTION IMAGING IN THE X-RAY

The goal of astronomy is to make the distant appear close, since the extreme distances of the universe obscure our view of its components and hide the workings of nature. Through the use of telescopes astronomers have improved our vision, and physical understanding of the universe has followed.

The Hubble Space Telescope represents the greatest clarity of vision ever achieved by a major observatory at visible wavelengths. The 0.1 arcseconds resolution it achieved is only six hundred times finer than that experienced with the naked eye. But, the results have been stunning anyway.

Intercontinental baseline radio interferometry has produced images with milli-arcsecond resolution, one hundred times finer than HST. With these images, astronomers have probed deep into the hearts of quasars and the Milky Way Galaxy, but are limited to highly non-thermal sources.

Astronomers are nowhere near reaching the practical limits of imaging. Many orders of magnitude improvement are possible across much of the electromagnetic spectrum.

To achieve higher resolution, the astronomer must move to larger aperture telescopes or longer baseline interferometers to beat down the effects of diffraction. However, as the quality of the image improves, the telescope must be built large enough to collect an adequate signal. Thus the three basic parameters for improving imaging are wavelength, baseline, and collecting area.

But what is the practical limitation? Space scientists have shown that cosmic objects may be observed at any wavelength. The diameter of a telescope determines its resolution until it becomes so large that it is broken into separate telescopes. At that point the resolution is determined by the baseline. Baseline costs very little to increase once one is combining telescope beams. On the other hand, collecting area always costs money. At the very least, costs rise linearly with collecting area. All astronomers find themselves building the largest telescope they can afford, so they can observe the faintest, most distant possible objects.

The maximum affordable collecting area is somewhat a function of the band for which it is built. However, most large telescopes cannot exceed about one million square centimeters of collecting area, and observations of 10,000 seconds are typical for a multi-use observatory. Thus a grasp of $10^{10} cm^2 s$ is near the practical limit financially.

Consider that most sources of interest are thermal. With certain exceptions (like extreme synchrotron sources in the radio and lasers in the visible) the brightest sources at any given temperature are thermal blackbodies. Thus they emit with a brightness of $1.8 \times 10^{-5} T^4$ erg/cm^2/s/ster. By Conservation of Brightness, this is the same brightness we see at Earth.

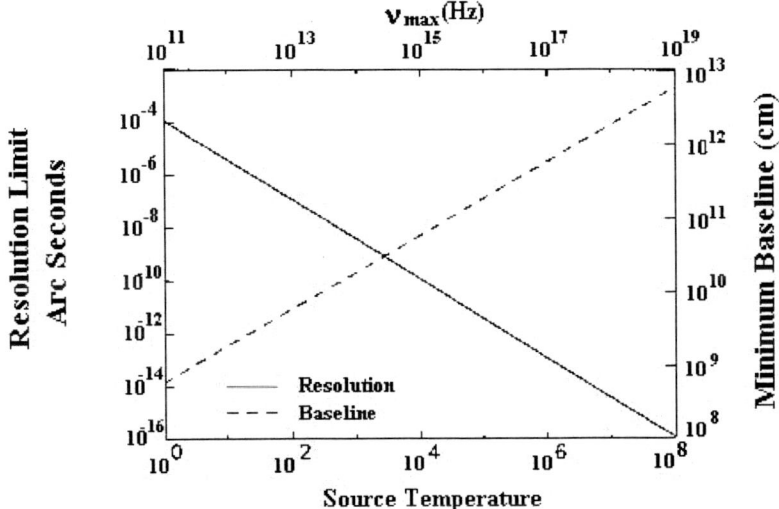

FIGURE 1. At plot of the minimum detectable angular blackbody feature with a reasonable size telescope as a function of temperature. Also shown is the baseline needed to resolve that feature. It is clear that the x-ray band provides the greatest resolution with only modest increases in baseline.

The brightness at Earth of an optically thick object rises as the fourth power of its temperature. Even adjusting for the fact that the photons become more energetic, the photon flux rises as the third power of the temperature. Thus a 5 million degree blackbody is a trillion times brighter than a 5 thousand degree object the same size. It emits just a billion times more photons.

X-ray astronomy has a reputation for dim sources. However, these fall into two classes. Some of the sources, like supernova remnants, are highly optically thin, and thus have low surface brightness. Others, like the inner parts of accretion disks or the surface of neutron stars, are optically thick (or nearly so). This makes them among the brightest sources in the universe, in the true sense of the word brightness. The reason that x-ray sources are faint is that they are small. However, that makes them ideal targets for super-high resolution imaging.

The peak frequency of emission from a blackbody of temperature T is given by:

$$v_{max} = 10^{11} T \qquad (1)$$

The number of photons detected in an instrument is given by:

$$N = \frac{1.8 \times 10^{-5}}{h v_{max}} G \theta^2 T^4 \qquad (2)$$

Where N is in photons/cm^2/s/ster.
Requiring N to be 100 photons and grasp G to be 10^{10} cm^{2}s, we find:

$$\theta_{min} = 6 \times 10^{-10} T^{-1.5} \qquad (3)$$

where θ_{min} is the minimum detectable angular feature in radians.

The baseline required to resolve θ_{min} is:

$$L = \frac{c}{v_{max} \theta_{min}} = 5.8 \times 10^8 T^{1/2} \qquad (4)$$

where L is the baseline in centimeters

Assuming a grasp of 10^{10} cm^{2}s, and a requirement of 100 photons detected per resolution element, we find that the minimum detectable feature size (θ_{min}), scales as $T^{-1.5}$. Such a strong function of temperature indicates that ultra-high resolution imaging is more tractable at high temperature and high energy.

In Figure 1 we show this effect graphically. As the temperature of an object rises, the minimum detectable feature size drops dramatically, while the required baseline rises only slowly. For example, in M87, at a distance of 15Mpc, the smallest feature size detectable in the visible would be 1.7×10^{-15} radians in extent, or 7.6×10^{10} cm, about the radius of the Sun. However, in the x-ray, the angular limit would be 5×10^{-20} radians, or about 22km! The baseline for the visible observation would have to be around 400,000km, the distance from the Earth to the Moon. The baseline for the vastly more powerful x-ray image would be larger, about 12 million km.

Of course, x-rays images differ greatly from visible images. X-rays are emitted only under conditions considered extreme by humans. Temperatures of millions of degrees and magnetic fields of millions of Gauss can create x-rays. They are often associated with the dramatic events heralding the both the birth and death of astronomical objects. As such, they come from compact regions and image the core structures in some of the most interesting events in the universe. This is the antithesis of structures viewed in radio VLBI, which are usually created by high energy electrons expanding away from the central structure. With x-rays we see the central engine itself.

A fundamental limitation in the clarity achieved by telescopes is to be found in the diffraction limit. As we all learned in school, the resolution limit of a telescope (expressed in radians) is λ/D, where λ is the wavelength of the observed radiation, and D is the diameter of the telescope aperture.

$$R = \frac{\lambda}{36000D} \quad (5)$$

The resolution R of a telescope (in arcseconds) is given, where λ is in Angstroms, and D is in meters. Example: HST has 2.5 meter diameter and wavelength of 5000Å, for a resolution of 0.055". In Equation 5 we convert this equation to more convenient units.

VLBI uses a wavelength of 2×10^8 Å on a baseline of 10^7 m to achieve .001 arcseconds. HST, with a 2.4m aperture at 5000Å achieves 0.1". The planned Space Interferometry Mission (SIM) uses 5000Å on a 20m baseline to achieve resolution of .01". In the x-ray where wavelengths can be as short as 2Å, it takes a one millimeter aperture to match HST, a one centimeter aperture to match SIM, and a full 10cm aperture to match intercontinental baseline interferometry in the radio. Truly, the diffraction limit is a much smaller problem in the x-ray, if optics of appropriate quality can be built.

So why do x-ray astronomers live and work with among the poorest quality images of any spectral band? The Chandra Observatory represents the extreme state of the art in x-ray observatories. Its resolution of one arcsecond, and collecting area of a thousand square centimeters can be matched in the visible portion of the spectrum by a mail order telescope selling for under $1000. The problem is that x-ray optics have been very difficult to fabricate to high quality. Wolter optics, with their quasi-cylindrical surfaces, are very hard to figure and polish.

Scientific Potential

Stellar Systems: The Sun's nearest neighbor, Alpha Centauri, is 260,000 times farther from the Earth as is the Sun. Thus, if we can improve our clarity of vision by a million times, we can observe our neighboring stellar systems with the same resolution we now enjoy within the solar system. It is well known that most stars have active, stellar cycles like our Sun. They emit copious x-rays from their corona and generate stellar winds, many of which are greatly more intense, as the Sun was billions of years ago. Such activity makes these systems observable in the x-ray.

At one micro-arcsecond, the disks of the stars will not only be resolved, but will present high quality images allowing the study of plasma dynamics, and greatly

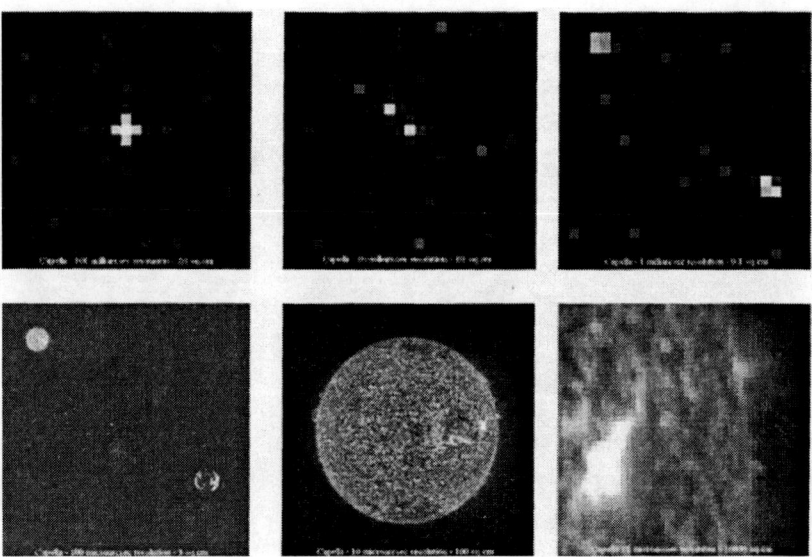

Figure 2: Simulations of x-rays from Capella (14pc away) are shown with differing resolutions. At the upper left, 0.1 arcseconds shows a point source. Successive orders of magnitude improvement in resolution reveal the binary nature. With 1 micro-arcsecond resolution (lower right), we are studying details of the surface.

improving our understanding of solar and stellar activity. With 500km resolution, the magnetospheres of the planets can be mapped. Fluorescence from the surfaces of the planets, moons, and asteroids can be detected and analyzed for composition.

Figure 2 is a sequence of simulated observations of the star Capella at increasingly high resolution. We see that higher resolution is, in some sense, the equivalent of visiting the star. As we approach the star, first we see it as a binary, then start to resolve the disks of the two components. Finally, we home in on the surface of one of the two stars. An x-ray interferometer would allow us to map and study many stars, and parts of their solar systems.

Event Horizons The event horizon of a black hole is the surface within which light cannot escape the intense gravitational field of the hole itself. Its existence is one of the most fascinating predictions of Einstein's theory of general relativity. Indeed, observations of physical phenomena occurring in the close vicinity of an event horizon would allow us to dramatically extend our understanding of gravitational phenomena under the most extreme conditions.

Our ultimate goal, of course, would be to take a "picture" of an event horizon. As of yet, this has been well beyond imaging capability in all wavebands, and especially in the higher energy bands (X-ray and gamma-ray) where radiation from matter entering the event horizon is preferentially emitted. However, the advent of X-ray interferometry could enable such a project. The best choice of target is the center of our own galaxy, the Milky Way, which is believed to harbor a black hole with a mass of 2.6×10^6 solar masses [1]. For a Schwarzschild black hole, the radius of the event horizon is 7.8×10^{11} cm, which, at a distance of 8kpc, subtends an angle of 6.6 microarcseconds. The galactic center is a known compact X-ray source, amenable to such an observation.

We can only imagine what such an image might display. Some of the earliest theoretical studies of the characteristics of radiation emitted close to the event horizon of both Schwarzchild and Kerr black holes (2) suggest a multitude of physical effects including strong gravitational lensing, Doppler and gravitational wavelength shifts, tidal distortions, dragging of inertial frames, and extreme time dilation, which can produce clear, unmistakable, spatial, spectral, and temporal signatures in the emitted X-ray light. To quote from Cunningham and Bardeen (2): "The effects we have calculated are quite striking, and if they could actually be observed, they would provide detailed, unambiguous information on strong-field predictions of general relativity."

Supernova 1987A: Perhaps the most timely of the possible future observations is to be found in the LMC. The blastwave from the supernova is just starting to impact the interstellar ring, and the surface brightness should increase by over a factor of 100. X-ray imaging of the interaction between the blastwave and the circumstellar ring surrounding the supernova explosion will certainly be spectacular. HST observations by Burrows et al. (3) show a triple ring structure consisting of material illuminated and then superheated by the initial radiation from the supernova blast. Because the ring is not uniform or symmetric (4) the x-ray luminosity is a highly sensitive function of density, the ring will quite literally sparkle. Imaging spectroscopy with STIS is already showing some small features on the interior of the ring growing in brightness (5)

Since the shocked gas is at temperatures $\geq 10^6$ K, x-ray observations will provide the most direct and least ambiguous information about the shock dynamics. Interstellar shocks play a critical role in astrophysics and, up to now, have only been studied in "snapshots" (with the notable exception of the impact of comet Shoemaker-Levy on Jupiter). With SN1987A, a small high resolution x-ray telescope would probe the subtle hydrodynamics of multiple shock reflections by supplying a movie of their behavior. Because these shock dynamics will unfold on a significantly subarcsecond scale, not even Chandra will be able to provide the community with clear images. Obtaining multiple x-ray images at the same resolution as HST will prove to be a critical element in understanding the dynamics of this unique object.

Other Supernova Remnants: The X-ray examination of SN1987A will provide great insight into the behavior of supernova shockwaves as they impact the surrounding interstellar medium, but we can gain substantial information about supernova remnant (SNR) physics at longer time scales by observing other known SNR's in our galaxy. The hot gas in these "older" SNR's is very highly ionized, emitting few optical or UV lines but copious soft X-rays. X-ray observations provide direct information on the morphology, chemical composition and evolution of SNR's beyond the first few years. Getting an accurate handle on the amount of gas in these objects as well as their structure and interactions with the ISM requires high resolution x-ray images.

One possible target is the Vela SNR which not only has a range of different structures seen in the X-ray, but also appears to have multiple "blobs" of ejecta traveling ahead of the supernova blastwave (6,7). Imaging these blobs and the associated shock-fronts at high resolution would tell us much about SNR development as well as help to refine hydrodynamic modeling. Another SNR of interest would be Cassiopeia A, which has 12,000 km/sec jets along with highly red and blue shifted x-ray emission lines coming from ejecta traveling at velocities of up to several thousand km/sec (8). These observations would resolve the suspected knots in these jets and identify the optical counterparts to the ejecta, giving the scientific community a bounty of new and valuable information.

Planetary Observations: Jovian and Solar system astronomy will also be enhanced. Among the areas of interest will be giant planet aurora; interactions between Jupiter, Io, and the Io flux tube; and lunar x-ray fluorescence. Jovian auroral observations provide a particularly exciting example of the power of high resolution optics. The Jovian x-ray aurora characteristically dissipates 10^{10} Watts of power, and shows complex luminosity and hardness ratio variations over several relevant timescales in the Jupiter/Io system. ROSAT, though limited to 3" resolution, has shown evidence for variable spatial structure, which probably results from variations in the input magnetospheric electron or heavy ion populations (9).

X-ray Binaries and SS433: X-ray binaries belong to an interesting class of objects wherein a compact object orbits a large star whose outer envelope is overflowing its Roche lobe. This material pours onto the compact object and in the process emits x-rays. Since the angular extent of the accretion disks in galactic binaries is on the order of micro-arcseconds, an image of a mass transfer binary will feature a point source surrounded by a scattering halo from interstellar dust.

One of the most curious mass transfer binaries is SS433. Discovered as a bright source in the optical, radio and x-ray, this massive binary system has twin opposed jets expelling material at 0.258c (10). What makes SS433 so exciting is that these jets and the mechanisms that formed them are thought to be similar to those found in quasars (11). SS433 can be thought of as a nearby testbed for quasar theories. The jets have been resolved in the radio (12) but no x-ray instrument has yet had the required resolution of <0.15 arcsec (13) to resolve them.

Quasars and AGN's: Certainly one of the most immediate and exciting applications of higher resolution is the observation of active galactic nuclei (AGN's). AGN's are well known for their jet structures, seen in the optical and radio. There are hints of these structures occurring in the x-ray, but no instrument has been able to more than marginally resolve them. Whether or not there is a true x-ray jet is very important to the understanding of the central engines of these powerful objects, particularly since the x-rays emanate from the hottest and most energetic regions.

Jets are not the only interesting x-ray aspects of AGN's. There are a range of issues involving what is known as the Unified Model of AGN's (14). In particular, determining the physical extent of the central x-ray emission region as a function of AGN sub-class would be a major scientific advance, contributing to the verification or modification of the unified model. It may also be possible to resolve the region in which the outflow interacts with the narrow line region clouds. Eventually x-ray interferometry can give us direct images of the engine at the center.

Clusters of Galaxies: One of the most interesting recent astronomical x-ray observations has been the identification of hot x-ray emitting gas in clusters of galaxies. Examination has indicated the presence of cooling flows, wherein the hot gas appears to be collapsing towards the core of these clusters as a result of x-ray cooling. X-ray images show a characteristic intensity peak in a core cooling region of the cluster. The x-ray intensity profile can be used to determine the rate at which mass collapses into the core region due to cooling and, with some assumptions about cluster symmetry, can also generate a mass profile for the cluster (15). The mass rate at which a cluster cools is a critical number for determining how galaxies form and which galaxies form first. The mass profile returns essential information concerning the nature and distribution of dark matter, and thereby directly influences our estimation of the mass of the universe. Better x-ray resolution will permit the above determinations in a wider range of clusters as well as dramatically improving the accuracy of the values derived from current instruments.

The study of x-ray emission lines from clusters has indicated that the plasma is in a multiphase state, implying the presence of spatial structure (16). Only in the past few years has structure been resolved from any cluster (in A2029 (17) and in 2A 033+096 (18)). The presence of this structure in A2029 was, however, not confirmed by later ROSAT observations (19). The nature of this structure is one of the major unresolved questions surrounding cooling flows and has substantial implications.

It is possible that this structure derives from initial inhomogeneities in the mass distribution of the cluster. Such density variations would lead to enhanced cooling in small subregions and their subsequent collapse. This would have the observational consequence of forming knots or clumps on scales as small as 10 pc (20). It is also possible that this structure is due to an absorbing gas (whose presence has been hinted at by x-ray spectra) with a low covering factor (21). This cool gas might be where the x-ray gas goes once it has "cooled". In any case, these questions and the many more which have been raised by the discovery of cooling flow structure can be resolved through higher resolution x-ray imaging.

The Technical Challenge

First Achieve the Diffraction Limit

There is no point in building an interferometer until one has reached the practical limit on size for a single optic. Radio interferometry started after Jodrell Bank approached a feasible size limit. Visible light interferometry is necessary if one is to attain substantial advantage over 10m class mirrors. Thus 100m class baselines are under discussion.

In the x-ray, where we must work at grazing incidence, the limit is reached when the mirrors become too long. 30cm long mirrors at grazing incidence, with several in sequence are highly practical. It is not difficult to imagine a chain of one-meter mirrors. A chain of 3 meter mirrors is clearly pushing the edge of the envelope.

Figure 3: The first spherical mirror telescope is shown. It is composed of four 30cm spheres in sequence and was used to acquire the image shown in Figure 4.

A 3m mirror at 2 degrees of graze yields a 10cm aperture. 10Å radiation entering a diffraction limited optic with a 10cm aperture experiences diffraction at the 2 milli-arcsecond level. The optic would have to maintain somewhat better figure than a diffraction limited visible telescope. The wavelength is one thousand times shorter, but the graze angle buys back a factor of thirty. Thus, where one might specify a $\lambda/4$ optic in the visible, one will need a $\lambda/120$ optic for the x-ray device. Such tolerances are challenging, but well within the current state of the optical art.

Thus, if high quality mirrors suitable for x-ray astronomy can be built, the diffraction limit can be reached. We have recently developed and demonstrated a technique that allows us to build x-ray optics to exacting tolerance. This idea (22) involves using conventional, normal incidence mirrors at grazing incidence. It turns out that it can be shown that the primary aberrations can be removed by a sequence of four mirrors. Similarly, magnification of the focal plane is also possible. Since the optics required are no longer extreme aspheres, they are much cheaper and of much higher quality than Wolter class optics. Extensive raytracing shows that this design class can nicely support interferometry. The single telescope can reach 20mas (milli-arcseconds) for a 30cm optic, rising to 2mas for 3m optics. Thus, these telescopes are providing a breakthrough in their own right.

We know for a fact that the approach works and will be able to adapt it because, we have built and tested a four-element system that approached the diffraction limit. (see

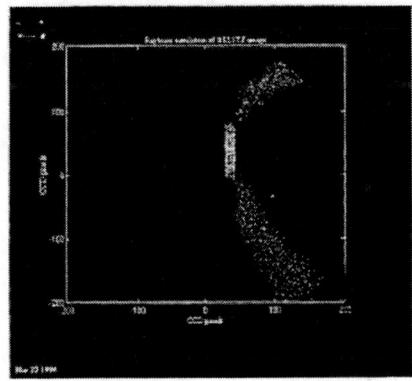 

Figure 4: Parallel bars were imaged at 13.3Å using the telescope in Fig 3. To the left is a simulation of what was expected. To the right is the actual image. Resolution near 0.1" was reached.

Figures 3 and 4)

Our single greatest problem in the search for high resolution imaging lies in the process of polishing a Wolter Telescope. Polishing any extreme asphere is exceedingly difficult, as the polishing lap is not free to to rotate and translate at will, as it is for a spherical mirror. This property makes aspheres more expensive and lower quality than spheres. With this in mind, we have been developing the use of spherical surface mirrors (or more generally normal incidence optics) used at grazing incidence. The telescopes then require two reflections per dimension to achieve a high resolution focus, as shown in Figure 4. We have discussed this approach to optics and provide the mathematical basis for the design class in patent 5,604,782. The patent also covers the generalization to interferometry. We have performed extensive tests that verify the large improvement in quality at a lower cost. Gallagher et al (23) report the first laboratory version of the spherical mirror telescope. Our newest results show good modulation all the way down to 0.1 arcseconds in our test telescope. NASA is supporting its application in building a sounding rocket to image x-rays at the 0.1" level.

The use of normal incidence optics at grazing incidence has now shown that we can achieve diffraction limited resolution in the x-ray. This will carry us another two or three magnitudes higher in resolution (close to one milli-arcsecond) before diffraction is likely to become a barrier. Moving beyond that point will require interferometry. There is more than one possible approach to interferometry, but the one I find most attractive is the Michelson Stellar Interferometer approach (24,25). It is particularly attractive in its simplicity. Two beams from co-aligned telescope apertures are mixed in the focal plane - nothing more. This is the approach in use for visible light interferometers, and it is directly applicable in the x-ray as well. Inasmuch as studies of the Space Interferometry Mission (26), show that astrometry is possible at the micro-arcsecond level, the same tolerances applied to an x-ray telescope would provide x-ray imaging at the micro-arcsecond level. While one milli-arcsecond requires a 100 meter baseline in the optical, it takes only 10 centimeters in the x-ray. One micro-arcsecond requires a 100km baseline in the visible, totally impractical for a single spacecraft, but requires only 100 meters in the x-ray. This is one application in which the very short wavelength of the x-ray is actually advantageous. Thus, for the first time it becomes possible to entertain the concept of imaging the coronae of other stars, and taking a direct image of the Event Horizon around a Black Hole.

Mix Two Beams

To achieve the synthetic aperture that is the goal of interferometry we must mix the beams from two grazing incidence telescopes in the focal plane (Figure 5). Each beam should be operating at the diffraction limit, and fringes will appear in the focal plane, in the classical, two slit Fraunhoffer way. To observe the fringes, however, one must have monochromatic light, and pathlengths between the two telescopes that are close

Schematic of Interferometer

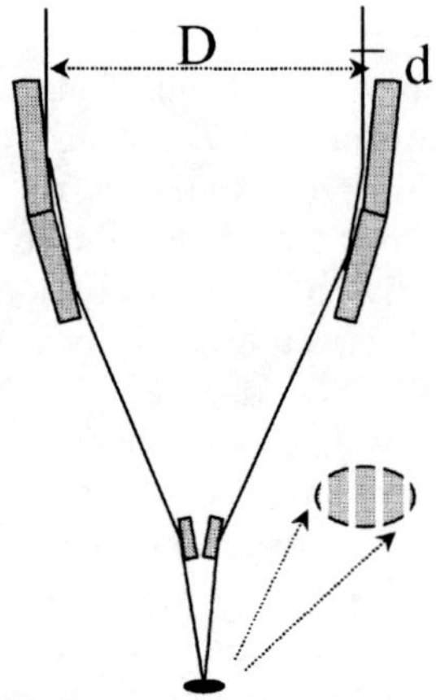

Figure 5: The beams from two independent x-ray telescopes can be mixed to form fringes.

to equal.

Fringes from different wavelengths have different frequencies and will thus wash out if allowed to mix. Similarly, as the optical pathlength difference to the focal plane grows, the sensitivity to small differences in wavelength grows, making the monochromaticity requirement even tighter.

To solve the problem of monochromatic light when the sources are panchromatic, there are basically three approaches. First, one can filter the signal from the star, keeping only a narrow band. In the x-ray this is typically accomplished with crystals or multilayer mirrors. However, the impact on the astronomy is devastating as the signal drops dramatically. Next best is to disperse the signal in one dimension according to wavelength, and have fringes in the other dimension. This is more acceptable, but still requires a substantial loss in signal (3 to 10 times) in the dispersing element. Best of all is to use the energy resolution of the detector. A CCD has $E/\delta E$ of nearly 20 at 1keV, supporting 20 fringes across the diffracted image. A hypothetical imaging quantum calorimeter with 1eV resolution could support a thousand fringes.

Improvements Shift To Mechanical

Once one has built two diffraction limited telescopes and mixed the beams to create fringes, improvements no longer come from the optics. There is little incentive to improve the quality of the mirrors. Consider, for example, that the optical surfaces of the radio dishes used in intercontinental baseline interferometry are made of coarse, unpolished metal, and fall many orders of magnitude short of milli-arcsecond optical quality. *Once below the diffraction limit, the quality of the mirror does not need to improve.*

TABLE 1: X-ray Interferometer Tolerances

Resolution Arcseconds	0.1	10^{-2}	10^{-3}	10^{-4}	10^{-5}	10^{-6}	10^{-7}
Mirror Length (m)	0.1	0.3	3	3	3	3	3
Position Stability (μ)	20	2	0.2	0.2	0.2	0.2	0.2
Angular Stability (arcsec)	10	2	.3	.1	10^{-2}	10^{-3}	10^{-4}
Figure	$\lambda/20$	$\lambda/50$	$\lambda/100$	$\lambda/100$	$\lambda/100$	$\lambda/100$	$\lambda/100$
Polish (Å rms)	30	20	20	20	20	20	20
Baseline (m)				1	10	100	1000
Angular Knowledge (as)	.03	3×10^{-3}	3×10^{-4}	3×10^{-5}	3×10^{-6}	3×10^{-7}	3×10^{-8}
Position Knowledge (nm)				20	20	20	20
E/dE Detector				20	20	20	20

Improvements below the diffraction limit are driven more by mechanical considerations. Higher spatial frequencies are achieved by moving the entrance apertures farther apart. Then, as the resolution and structures both grow, the stability of mechanical structures and aspect knowledge also must improve. The spectral resolution of the detector must grow as well. However, mechanical structures are in much better shape than are optics, and several orders of magnitude become immediately available once we achieve the diffraction limit.

The Practical Limit

Every approach to telescope building has practical limits. Some can be solved with sufficient application of money. However, some are simply beyond the current abilities of technology. In Table 1 we present a tolerance analysis table that shows the situation for x-ray interferometry as we push to ever higher resolution.

We can avoid interferometry entirely by building longer telescopes until such time as we need diffraction limited optics in excess of 3m in length. To achieve 0.1mas, a 30m optic would be required, so it is at 0.1mas that, as shown in Table 1, we convert to interferometry.

As the resolution goal gets tighter, many of the requirements become more challenging. At one micro-arcsecond, the requirements are challenging, but can probably be met. At 100 nano-arcseconds, the kilometer baseline stands out as something unavoidable. To move beyond will require multiple spacecraft with station-keeping. Thus we see one micro-arcsecond as the goal for the foreseeable future.

Implications for Small Imaging Missions

Clearly, x-ray imaging has significant potential within the constraints of small missions. From our discussion of the limitations on imaging and the possibilities we can conclude:

- Since many x-ray targets are among the brightest objects in the sky, high resolution imaging of tiny features can be achieved with surprisingly small apertures.
- With telescopes such as the spherical optic chains we discuss, it is possible for small missions to directly image the sky at 10 milli-arcsecond resolution.
- With small x-ray interferometers, a mission can be configured to reach 100 micro-arcseconds.
- With a cluster of small spacecraft, maintained in a phase configured array, it is possible to achieve resolution of a micro-arcsecond or better.

ACKNOWLEDGEMENTS

I wish to thank R. McCray for discussions on high resolution imaging, and S. Sarlin, and J. Farmer for help with the x-ray optics. This work was supported by NASA grant NAG5-5020

REFERENCES

1. Eckart, A. and Genzel, R., *Bull.Amer.Astr.Soc.*, **191**, 97.07 (1997)
2. Cunningham, C. T., and Bardeen, J. M., *Ap. J.*, **183**, 237 (1973)
3. Burrows, C. J., et al., *Ap. J.*, **452**, 680 (1995)
4. Plait, P. C., Lundquist,P., Chevalier, R. A., Kirshner, R. P., *Ap. J.*, **439**, 730 (1995)
5. Sonneborn, G., et al., *Ap.J.(Letters)*, **492**, L139 (1998)

6. Aschenbach, B., Egger, R., Trumper, J., *Nature*, **373**, 587 (1995)
7. Strom, R., Johnston, H. M., Verbunt, F., Aschenbach, B., *Nature*, **373**, 590 (1995)
8. Fesen, R. A., Gunderson, K. S., *Ap. J.*, **470**, 967 (1996)
9. Waite J. H., et. al., *J. Geophys. Res.* **99**, 14799 (1993)
10. Margon, B., *Ann. Rev. Astron. Astro.* **22**, 507 (1984)
11. Lyutyi, V. M., Cherepashchuk, A. M., *Sov. Astron.* **30**, 532 (1986)
12. Hjellming, R. M., Johnston, K. J., *Ap. J.*, **328**, 600 (1988)
13. Fabrika, S. N., *MNRAS*, **261**, 241 (1993)
14. Antonucci, R., *Ann. Rev. Astron. Astro.,* **31**, 473 (1993)
15. Sarazin, C. L., "X-ray Emission from Clusters of Galaxies", Cambridge University Press (1992)
16. Canazares, C. R., Markert, T. H., Donahue, M., E., "Cooling Flows in Clusters and Galaxies", ed. Fabian, Kluwer Academic, p. 63 (1988)
17. Sarazin, C. L., O'Connell, R. W., McNamara, B. R., *Ap. J.(Letters)*, **389**, L59 (1992)
18. Sarazin, C. L., O'Connell, R. W., McNamara, B. R., *Ap. J.(Letters)*, **397**, L31 (1992)
19. White, D. A., Fabian, A. C., Allen, S. W., Edge, A. C., Crawford, C. S., Johnstone, R. M., Stewart, G. C., Voges, W., *MNRAS*, **269**, 589 (1994)
20. Nulsen, P. E. J., *MNRAS*, **221**, 377 (1986)
21. Antonucci, R., Barvainis, R., *Astro. J.,* **107**, 448 (1994)
22. Cash, W., US patent #5,604,782 (1997)
23. Gallagher, D., Cash, W., Jelsma, S., Farmer, J., *Proc. Soc. Photo-Opt. Instr. Eng.* , **2805**, 121, (1996)
24. Michelson, A. A., *Ap. J.*, **51**, 257 (1920)
25. Cash, W., "X-ray Interferometers", *Proceedings of the International School of Space Science* (1995)
26. Marr, J., Saterios, D., Laskin, R., Unwin, S., Yu, J., Proceedings of the 16th IEEE Instrumentation and Measurement, ed. V. Piuri and M. Savino, IEEE Press, 1117 (1999)

The Intergalactic Medium and Soft X-ray Background

Renyue Cen*

*Princeton University Observatory[1]
†Princeton, NJ 08544

Abstract. I present an overview of some of the recent advances in our understanding of the distribution and evolution of the ordinary, baryonic matter in the universe. Two observations that strongly suggest that most of the baryons seen at high redshift ($z \geq 2$) have turned into some forms yet undetected at $z = 0$ are highlighted. With the aid of large-scale cosmological hydrodynamic simulations, it is shown that most of the baryons today are in a gaseous form with a temperature of $10^5 - 10^7$ Kelvin – the "warm/hot gas", shock heated during the gravitational collapse and formation of the large-scale structure at low redshift. Primarily line emissions from this warm/hot gas may account for a large fraction of the residual (after removal of identifiable discrete sources) soft X-ray background at $h\nu < 1.0$keV. How this warm/hot gas may be detected by the next generation of EUV and soft X-ray instruments is indicated. Detection or non-detection of this warm/hot gas will have profound implications for cosmology.

INTRODUCTION

It is well known that most of the matter in the universe is in some non-luminous, dark form, with the cold dark matter being the most popular choice (Peebles 1993). Not only that, most of the ordinary, baryonic matter, which altogether makes up about 10-20% of the total matter in the universe, seems to be missing in the present day universe. At redshift $z = 2 - 3$, the amount of gas contained in the Lyman alpha forest is (Rauch et al. 1997; Weinberg et al. 1997)

$$\Omega_{b,\text{Ly}\alpha}(z = 2 - 3) \geq 0.017h^{-2} = 0.035, \qquad (1)$$

where $\Omega_{b,\text{Ly}\alpha}(z = 2 - 3)$ is the baryonic density in units of the critical density extrapolated to $z = 0$ and a Hubble constant $h \equiv H_0/(100\text{km/s/Mpc}) = 0.70$ is adopted throughout. Independently, the observed light-element ratios (in particular, the deuterium to hydrogen ratio) in some carefully selected absorption line

---

[1] Present research is sponsored in part by the National Science Foundation and the National Aeronautics and Space Administration.

systems at $z = 2-3$, interpreted within the context of the standard nucleosynthesis theory, yield the total baryonic density (Burles & Tytler 1998)

$$\Omega_{b,D/H}(z = 2 - 3) = (0.019 \pm 0.001)h^{-2} = 0.039 \pm 0.002. \qquad (2)$$

The agreement between these two completely independent measurements is remarkable. But, at redshift zero, after summing over all well observed contributions, the baryonic density appears to be far (by a factor of three) below that indicated by equations (1) and (2) (e.g., Fukugita, Hogan, & Peebles 1997):

$$\Omega_b(z=0)|_{\text{seen}} = \Omega_* + \Omega_{HI} + \Omega_{H_2} + \Omega_{Xray,cl} \approx 0.0068 \le 0.011 \quad (2\sigma \text{ limit}). \qquad (3)$$

Thus, unless two independent errors have been made in the arguments that led to equations (1) and (2), there is a sharp decline of the amount of observed baryons from high redshift to the present day; i.e., most of the baryons in the present day universe are yet to be detected. Now, the evolution of universal baryons can be computed from standard initial conditions in the modern cosmological setting, using realistic large-scale hydrodynamic simulations. It is found that the large amount of baryons are *not missing* but *hidden* in an intergalactic gas with a temperature $10^5 < T < 10^7$ Kelvin - the "warm/hot gas" - at $z = 0$, which is difficult to detect (§2). §3 shows observable signatures of the warm/hot gas presents imprinted in the soft X-ray background. Several ways to detect this warm/hot gas are suggested in §4 and conclusions presented in §5.

EVOLUTION OF COSMIC BARYONS

In a series of model simulations of nearly a dozen different models covering the current interest we have shown, consistently and robustly, that from 50% to 70% of all baryons in all models examined are shock heated during the gravitational collapse of the large-scale structure in the recent past and are in a warm/hot gas with a temperature of $10^5 - 10^7$ Kelvin at $z = 0$ (Ostriker & Cen 1996) with each model being approximately normalized to match the local large-scale structure. It was clearly and immdiately understood that this model independent outcome found in earlier work would have profound implications for cosmology and it is therefore of paramount importance to verify it with latest simulations that include more relevant physics (including feedback from star formation and metal cooling, etc.) and have more accurate treatment of shocks (Ryu et al. 1993) than our prior code (Cen 1992). Indeed, it was confirmed by a recent higher-resolution, larger-size hydrodynamic simulation of a cold dark matter model with a cosmological constant. The adopted model - the Ostriker-Steinhardt (1995) Concordance model - is normalized to both the microwave background temperature fluctuations measured by COBE on large scales (Bunn & White 1997) and the observed abundance of clusters of galaxies in the local universe (Cen 1998) with $\Omega_0 = 0.37$, $\Omega_b = 0.049$, $\Lambda_0 = 0.63$, $\sigma_8 = 0.80$, $h = 0.70$ and $n = 0.95$ (with 25% tensor contribution to the

CMB fluctuations on large scales). The simulation box is $L = 100h^{-1}$Mpc with 512^3 fluid elements and 256^3 dark matter particles. Three components are followed separately and simultaneously: dark matter, gas and "galaxies". The last component is created continuously (like real galaxies) during the simulation in places where local physical conditions permit rapid cooling and collapse, the dynamics of the aftermath of which cannot be followed in the present simulation. Instead, we allow star formation to occur in these regions where gravitational collapse cannot be reversed until stellar systems are formed, under plausible assumptions. In addition to standard physics in cosmological gasdynamic simulations, feedback into the intergalactic medium (IGM) from star formation is allowed in three related forms: UV radiation, supernova energy and mass ejection. Cooling due to metals is also included. The model reproduces the observed evolution of the luminosity density of the universe at various energy bands (Nagamine et al. 1999), the evolution of galaxy clustering (Cen & Ostriker 1998) and metallicity distributions (Cen & Ostriker 1999b) among others.

The results from this new simulation focusing on the evolution of the cosmic gas have been presented in Cen & Ostriker (1999a) and are summarized here. We divide the baryonic gas into three temperature ranges (1) $T > 10^7$ K (the X-ray emitting gas in collapsed and virialized clusters of galaxies); (2) 10^7 K$> T > 10^5$ K gas, which we will call the warm/hot gas and is located outside of clusters of galaxies; (3) $T < 10^5$ K warm gas, which is seen in optical studies as Lyα clouds or Gunn-Peterson effect and is primarily in voids at $z = 0$. A last component (4) is the cold gas that has been condensed into stellar objects, which we designate "galaxies". Figure 1 shows the evolution of these four components, and the results are consistent with our other knowledge. Most of the baryonic mass is in warm gas (Lyman alpha forest) at $z = 3$ making up 94%, which declines with increasing time to 26% at $z = 0$, consistent with the HST observed clearing of the forest and of low-z redshift Lyα cloud gas (e.g., Shull 1996, 1997). The hot component increases in mass fraction with increasing time, reaching 12% at $z = 0$, and is consistent with observations of the local properties of the X-ray emitting great clusters (e.g., White et al. 1993; Cen & Ostriker 1994; Lubin et al. 1996; Bryan & Norman 1998). The condensed component remains small, consistent with the known mass density in galaxies (e.g., Fukugita et al. 1997).

Our attention will be focused on the solid circles in Figure 1: the warm/hot gas rises rapidly with increasing time and dominates the baryonic mass budget by $z = 0$, reaching 52% of the noncondensed mass fraction or 47% of the total baryons. Also shown in Figure 1 is the warm/hot component for two other models, an open CDM model with $\Omega_0 = 0.40$ and $\sigma_8 = 0.75$ (dotted curves), and a mixed hot and cold dark matter model with $\Omega_{hot} = 0.30$ and $\sigma_8 = 0.67$ (dashed curves) computed completely independently by Bryan & Norman (1998). Quite reassuringly, their results are in excellent agreement with ours. The density fluctuation amplitude normalization of their mixed dark matter model is somewhat below that required to produce the abundance of local galaxy clusters. Therefore, an appropriately normalized mixed dark matter model would yield a larger warm/hot gas fraction thus

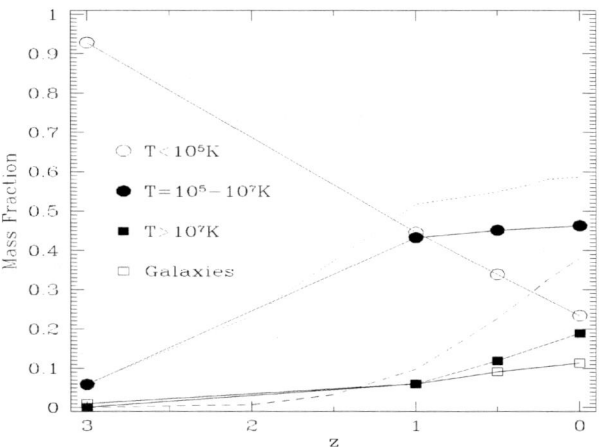

FIGURE 1. Mass Fractions of the four components (see text) as a function of redshift.

is in still better agreement with the other two models, re-enforcing the conclusion that the warm/hot gas makes up most of the baryonic matter today, independent of models as long as each model is normalized to match the local large-scale structure. Physically, this may be understood as follows. The temperature of the bulk of the gas today should be determined, to the zeroth order, by the velocity of converging waves that are collapsing today. The length of the waves that have become non-linear today is about $8h^{-1}$Mpc, which is almost exactly the scale that is used to normalize each model to match the local abundance of clusters of galaxies. It is worth stressing that the reason for most of the gas phase being in the warm/hot gas is primarily gravitational, as implied above. In other words, the gas is primarily shock heated during the gravitational collapse and formation of the present large-scale structure. Other potentially relevant physical processes, such as the meta-galactic radiation field, metal cooling and energy deposition into IGM from young galaxies, which were included in the simulation examined here, are shown to be not of primary importance (Cen & Ostriker 1999a), with increasing, secondary importance in that order.

Figure 2 shows the spatial distribution of this warm/hot gas (the box size is $100h^{-1}$ Mpc) at $z = 0$. Figure 3 shows the spatial distribution of hot cluster gas with $T > 10^7$ Kelvin at $z = 0$. We see in Figure 2 a filamentary network of the warm/hot gas with hot cluster gas (see Figure 3) residing at the intersections of filaments. These regions of warm/hot gas - groups of galaxies, filaments and sheets - typically have such a low surface brightness that current instrumentation does not detect them as distinct "sources". The typical size (width of the filaments) is about one to several megaparsecs, corresponding to an angle of order a degree or so, if placed at a distance of a few hundred megaparsecs. The lengths of these

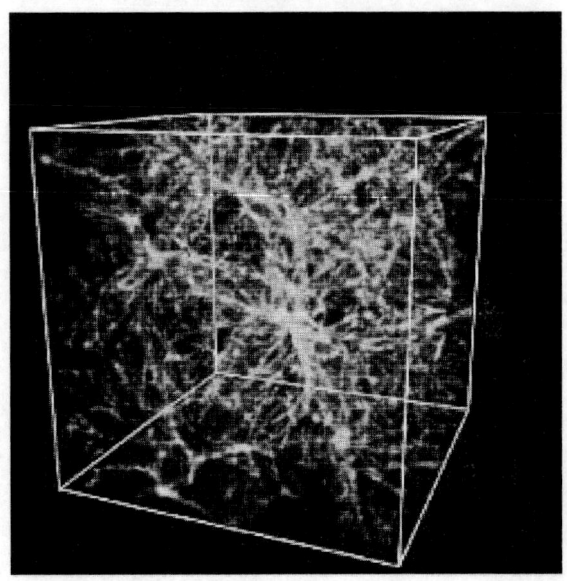

FIGURE 2. Spatial distribution of the "warm/hot gas" in a box of size $100h^{-1}$Mpc.

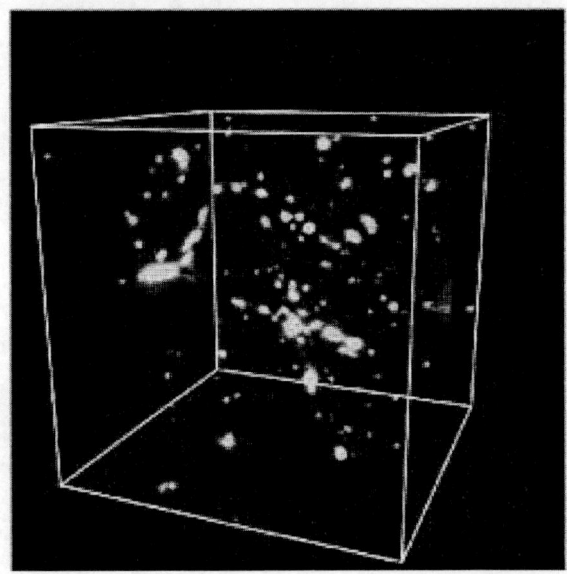

FIGURE 3. Spatial distribution of the hot intracluster gas ($T > 10^7$Kelvin) in a box of size $100h^{-1}$Mpc.

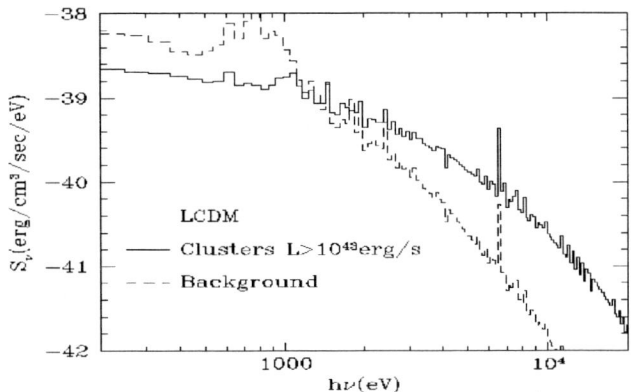

FIGURE 4. Emission from warm/hot gas and hot intracluster cluster gas as a function of frequency.

filaments are tens of megaparsecs. The typical density of the filaments is about $10 - 100$ times the mean density of the universe.

CONTRIBUTION TO THE SOFT X-RAY BACKGROUND FROM THE WARM/HOT GAS

It was shown in a simulation of a similar model but with somewhat lower resolution (Cen et al. 1995) that this warm/hot gas makes a nontrivial contribution to the soft X-ray background at < 1.0 keV. In the present higher resolution simulation we confirm this conclusion. After scaling Ω_b down from 0.049 to 0.037 using latest observations (Burles & Tytler 1998) and noting the $J \propto \Omega_b^2$ scaling relation, we find that 25% of the *total* extragalactic soft X-ray background at 0.7 keV comes from this diffuse warm/hot gas: $J_{WH} = 7$ keV/sec/cm^2/keV/sr. Most of the emission at < 1.0keV from the warm/hot gas is due to various blends of emission lines. Figure 4 shows X-ray emissivity of the background gas (i.e., warm/hot gas; dashed cruve) and of all the bright clusters with $L_{bol} > 10^{43}$erg/s (solid curve) at redshift $z = 0$. The primary spectral signature of the background in the region 0.5-1.0keV is the "iron bump" (a mixture of iron and oxygen lines, primarily). Major lines in this range include a blend of OVII lines at 0.561-0.574keV (characteristic of 10^6 K gas), a blend of lines from NeIX at 0.904-0.922keV (at a similar temperature), OVIII line at 0.654keV (peaked around $10^{6.4}$ Kelvin; the ratio of the 0.57keV to the 0.65keV features can be used as a temperature indicator) and a pair of iron XVII lines at 0.726keV and 0.739keV (which have peak strength at approximately $10^{6.5}$ K).

In the soft X-ray the Milky Way is a strong source of emission at an effective temperature of $\sim 10^6$ K. To some extent the soft X-ray background that we are proposing is the same as Galactic emission in that the sum over all galaxies like our own does make a nontrivial contribution to the soft X-ray background, since some of the warm/hot gas resides in the halos of galaxies. If the average galaxy emits 2×10^{39}erg/s in the 0.5-1.5keV band (cf. Fabbiano 1989), then the sum of all such will produce an emissivity of 10^{-39}erg/cm^3/sec/eV not far from the levels shown in Figure 4. But most of the emission shown by the dashed line in Figure 4 is due to hot gas in bigger systems than our own galactic disk or halo. Thus, the temperature (as weighted by ρ^2) is in the range $10^6 - 10^7$ K, whereas, for most of the galactic coronal gas, the typical temperature (as weighted by ρ^2) is in the range $10^5 - 10^6$ K. Thus, the line ratios indicative of the OVI/OVII and OVII/OVIII ratios will help in distinguishing between the two components. The spectral features which were noted above should help provide clues to the origin of the background radiation in this range as the galactic component is probably too cool to produce the iron blend which should be prominent in the background component described here. The steepening of the spectrum below 1keV seen recently by ASCA (cf. Gendreau et al. 1995) as well as the OVII lines (also noted by ASCA) are also tentative but direct observational evidence that the background gas we are discussing has already been detected. Recent ROSAT observations of the soft X-ray background (Wang & McCray 1993) seem to hint the existence of this warm/hot gas.

Shadowing by nearby neutral hydrogen-rich galaxies (McCammon & Sanders 1990; Wang & Ye 1996; Barber, Roberts, & Warwick 1996) quite convincingly shows that a significant fraction of at least the component in the range 0.5-1.0keV is truly extragalactic (cf. Burrow & Kraft 1993; Gendreau et al. 1995). Wang & Ye (1996) estimate that ≥ 4 keV/sec/cm^2/keV/sr out of the total soft X-ray background at 0.7 keV is truly diffuse in nature. It thus deos not seem difficult for the contribution from the warm/hot gas to account for the residual, diffuse soft X-ray background.

It is found from simulations that one half of the soft X-ray background due to the warm/hot gas is emitted by structures at redshift $z < 0.65$ and three quarters from $z < 1.0$. So one may be be able to identify the optical features associated with the emitting gas.

WAYS TO DETECT THE WARM/HOT GAS

The spectral features of this warm/hot gas are in the EUV and soft X-ray, which make it difficult to observe at low redshift due to possible confusion with features from the interstellar medium in our own galaxy. Proper identifications of spectral lines will allow to unambiguously distinguish local galactic features from extragalactic ones, for sufficiently distant, *individual filamentary* structures (perhaps $z \geq 0.01$). These individual structures may be detected directly in several ways. First, Hellsten et al. (1998), on the basis of similar simulations, have predicted the existence of an X-ray absorption forest due to ionized oxygen (O VII 574 eV line) in the warm/hot temperature range. Perna & Loeb (1998) have also made similar calculations based on simplified models for the IGM. Work underway with T. Tripp and E. Jenkins indicates that UV absorption lines (OVI 1032A,1038A doublets) due to gas in this temperature range primarily in the distant outskirts of galaxies may be detectable by current or planned instruments. Second, strong soft X-ray emission lines from highly ionized species (such as O VIII 653 eV line) should also be observable (Jahoda et al. 1998, in preparation). The next generation of UV to soft X-ray spectroscopic instruments with sufficient spectral resolution and sensitivity will provide direct ways to measure *"the X-ray forest"* or *"the X-ray large-scale structure"*, depending on whether it is seen in absorption or emission. Third, the warm/hot gas may show up as very broad, relatively weak (mostly having $N_{HI} \leq 10^{13} \text{cm}^{-2}$ with a small fraction at higher column densities), low redshift Lyα clouds (Shull 1996,1997; but observations more sensitive than current ones are required). NGST should be able to identify these local counterparts of the high redshift Lyα clouds.

If our prediction that most of the residual soft X-ray background is produced by the warm/hot gas is correct, then, since most of the soft X-ray background is produced by low redshift structures, this could be verified by associations of this soft X-ray background radiation field with relatively nearby large scale structure features, as well as by the soft X-ray angular auto-correlation function. A preliminary study (Cen et al. 1999, in preparation) indicates that the auto-correlation function of the soft X-ray background from warm/hot gas is positive up to a few degrees in separation, in good agreement with observations from ROSAT (Soltan et al. 1996). Cross-correlating soft X-ray background (Refregier, Helfand, & McMahon 1997) or Sunyaev-Zel'dovich effect (Refregier, Spergel, & Herbig 1998) with galaxies should provide additional important information. The next generation of X-ray instruments (AXAF and ABRIXAS), in conjunction with large-scale redshit surveys (e.g., Sloan Digital Sky Survey) should provide some potentially powerful tests of the existence of this warm/hot gas.

CONCLUSIONS

According to the current theory for the growth of cosmic structure, it is inevitable that most of the cosmic gas is shock heated during the course of gravitational collapse and formation of the present day large-scale structure and ends up in a warm/hot gas with a temperature $10^5 - 10^7$ Kelvin in the present day universe. This gas accounts for the so called missing baryons in our local universe, which constitutes most of the baryons today. This gas with density of $10 - 100$ times the mean density of the universe resides in a filamentary network with each individual filament having a length of tens of megaparsecs and a width of order one megaparsec, easily distinguishable from the much hotter gas in the centers of great clusters of galaxies, which is located in the intersections of the filaments and has a much higher temperature. The emission (primarily line emission) from this warm/hot gas dominates that from intracluster gas in the soft X-rays and may account for most of the residual soft X-ray background at < 1keV. This network of filamentary warm/hot gas is probably too faint to be detected as individual sources by available instruments. However, the next generation of instruments in EUV and soft X-ray should be able to detect this warm/hot gas, in a number of possible ways, suggested in the previous section. If this warm/hot gas is detected, a consistent picture of gravitational growth of structure will be affirmed once again; otherwise, the current theory of structure formation within the gravitational instability paradigm and/or of standard light element nucleosynthesis may require a re-examination.

The work presented here should be mostly credited to my collaborators, Drs. Jeremiah P. Ostriker, Hyesung Kang and Dongus Ryu. I thank Dr. Greg Bryan for allowing to use the results from his simulations before publication. I thank Dr. Steven Brumby for his warm hospitality and for organizing an entertaining and stimulating conference. The work is supported in part by grants AST9318185 and ASC9740300.

REFERENCES

1. Barber, C.R., Roberts, T.P., & Warwick, R.S., *M.N.R.A.S.*, **282**, 157 (1996).
2. Bryan, G.L., & Norman, M.L. *Ap. J.*, **495**, 80 (1998).
3. Bunn, E.F., & White, M., *Ap. J.*, **480**, 6 (1997).
4. Burles, S., & Tytler, D. *Ap. J.*, **499**, 699 (1998).
5. Burrows, D.N., & Kraft, R.P., *Ap.J.*, **411**, 685 (1993).
6. Cen, R., *Ap. J. Suppl.*, **78**, 341 (1992).
7. Cen, R., *Ap. J.*, **509**, 494 (1998).
8. Cen, R., Kang, H., Ostriker, J.P., & Ryu, D., *Ap. J.*, **451**, 436 (1995).
9. Cen, R., & Ostriker, J. P., *Ap. J.*, **429**, 4 (1994).
10. Cen, R., & Ostriker, J. P., **preprint**, astro-ph/9809370 (1998).
11. Cen, R., & Ostriker, J. P., *Ap. J.*, **514**, 1 (1999a).
12. Cen, R., & Ostriker, J. P., *Ap. J.*, **in press** (1999b).
13. Fabbiano, G., *A.R.A.A.*, **27**, 89 (1989).
14. Fukugita, M., Hogan, C.J., & Peebles, P.J.E., *Ap. J.*, **503**, 518 (1998).
15. Hellsten, U., Gnedin, N.Y., & Miralda-Escudé, J., *Ap. J.*, **509**, 56 (1998).
16. Lubin, L., Cen, R., Bahcall, N.A., & Ostriker, J.P., *Ap. J.*, **460**, 10 (1996).
17. McCammon, D., & Sanders, W.T., *A.R.A.A.*, **28**, 657 (1990).
18. Nagamine, K., Cen, R., & Ostriker, J.P., *Ap. J.*, **submitted** (1999).
19. Ostriker, J. P., & Cen, R., *Ap. J.*, **464**, 27 (1996).
20. Ostriker, J. P., & Steinhardt, P., *Nature*, **377**, 600 (1995).
21. Peebles, P.J.E., *Principles of Physical Cosmology* (Princeton University Press: Princeton, NJ) (1993).
22. Perna, R., & Loeb, A., *Ap. J.*, **503**, L103 (1998).

23. Rauch, M., Miralda-Escudé, J., Sargent, W.L.W., Barlow, T. A., Weinberg, D.H., Hernquist, L., Katz, N., Cen, R., Ostriker, J.P., *Ap. J.*, **489**, 1 (1998).
24. Refregier, A., Helfand, D., McMahon, R.G., *Ap. J.*, **479**, L93 (1997).
25. Refregier, A., Spergel, D.N., & Herbig, T., **preprint**, astro-ph/9806349 (1998).
26. Ryu, D., Ostriker, J.P., Kang, H., & Cen, R., *Ap. J.*, **414**, 1 (1993).
27. Soltan, A.M., Hasinger, G., Egger, R., Snowden, S., & Truemper, J., *A.& A.*, **305**, 17 (1996).
28. Shull, J.M., *A. J.*, **111**, 72 (1996).
29. Shull, J.M., in "structure and evolution of the IGM from QSO absorption lines", ed. P. Petitjean & S. Charlot (1997).
30. Gendreau, K.C., et al. , *P.A.S.P.*, **47**, 5 (1995).
31. Wang, Q.D., & McCray, R., *Ap. J.*, **409**, L37 (1993).
32. Wang, Q.D., & Ye, T., *New Astron.*, **1**, 245 (1996).
33. Weinberg, D.H., Miralda-Escudé, J., Hernquist, L., Katz, N., *Ap. J.*, **490**, 564 (1997).
34. White, S.D.M., Navarro, J.F., Evrard, A.E., & Frenk, C.S., *Nature*, **366**, 6465 (1993).

Perspectives of Astrophysical and Gravitational Research Onboard Small Spacecraft

Victor I. Denisov*, Sergey I. Svertilov*, Michail I. Kudryavtsev *
and Zinaida P. Cheryomukhina †, Vladimir B. Pinchoock †

*Applied and Theoretical Research Institute, Moscow State University, Russia.
† TSNIIMASH, Russia

Abstract. The physical properties of hard-radiation emitting astrophysical objects (X-ray binaries pulsars, active galactic nuclei) can be studied in detail using compact instruments on-board small spacecraft. The polarization and spectroscopic measurements are of particular interest. The search and study of emission lines in hard radiation of X-ray binaries, black hole candidates as well as in diffuse gamma-radiation from Galactic Center and star formation regions, are the main goals of spectroscopic observations. It seems that for further progress of spectrometric measurements, detectors based on pure or enriched germanium should be used. As for polarization measurements, searching and measurements of polarization of hard radiation of pulsars and black hole candidates are of great interest. If the exposure time of the same X-ray source will be about several years, it allows us to search the electromagnetic radiation component, caused by the interaction of high-frequency gravitational radiation of the neutron star with its magnetic field. Such observational data will be unique because until the present time continuous long measurements of X-ray polarization and its time variations have not been performed. These data allow to solve some problems of fundamental physics of the X-ray binaries and to estimate the upper limit of high frequency gravitational radiation intensity.

INTRODUCTION

Despite the great achievements of X- and gamma-ray astronomy there are no complete and comprehensive models of most astrophysical objects emitting hard electromagnetic radiation, such as X-ray binaries, pulsars and active galactic nuclei. Valuable information about such phenomena can be obtained by detailed measurements of energy spectra and polarization of X- and gamma-rays, that also allows us to study physical processes in the strong electromagnetic and gravitational fields near compact objects. Corresponding observations can be realized using rather small instruments on-board light and consequently cheap spacecraft.

Here we discuss in details the spectral and polarization measurements of hard radiation of compact astrophysical objects.

SPECTRAL MEASUREMENTS

The main goal of spectral observations in high energy astrophysics is the search and study of emission lines in hard X-rays and gamma-rays. Such lines were observed in spectra of several sources. In particular, the 511 keV annihilation line was observed in the spectra of several X-ray Novaes and X-ray binaries Cyg X-1. Such type of binaries are the most probable candidates in include black holes, thus the presence of annihilation line in the energy spectrum can be considered as the indicator of a black hole. The gamma-ray lines caused by some radioactive isotopes: ^{26}Al (1.8 eV), ^{56}Co was observed in diffuse Galactic radiation as well as from few molecular clouds and gas-dust complexes which are well-known as active star-formation regions [3 - 5]. Emission lines were observed in spectra of some other sources, in particular SS433 [6], and the supernova remnant SN1987 [7].

At the present time, spectral measurements in X- and gamma- ranges were realized mainly by the use of scintillation detectors, which have energy resolution in the range of 10-30%. Such resolution makes it impossible to measure intrinsic line profiles because the measurable profiles are caused by instrumental response function in this case. The further progress of spectral observations may be possible if detectors of high resolution, e.g., of pure enriched Germanium, will be used. It is the energy resolution of Germanium detectors $\approx 0.2\%$ at 1 V, that allows us to measure intrinsic line profiles. Background can be eliminated effectively if observation "in lines" will be realized with detectors of high resolution [8].

There are only few successfully completed gamma-astronomy experiments with germanium detectors: HEAO3, WIND. Of modern projects it is necessary to note INTEGRAL and Spectrum-Roentgen-Gamma as the most famous and advanced. Both mentioned above experiments and projects were realized or there are proposed to be realized on-board massive spacecraft. Background on such spacecraft is caused mainly by its own radiation (local and induced) which can lead to sufficient instrumental background even for high-resolution germanium detectors. Thus for spectral measurements it should be preferable to realize the experiment with germanium detector on-board super-light spacecraft able to carry payload with mass of 50-100 kg. Such spacecraft have some advantages in comparison with traditional space observatories. Due to the low mass of a spacecraft the intensity of the local gamma-quanta will be negligible and gamma-quanta background will be caused mainly by cosmic diffuse gamma-rays. Also the cooling of germanium detector to temperatures less than 100 K can be realized more simply on such spacecraft. For example, the temperature necessary for cooling of detector with effective area ≈ 40 cm^2, can be provided by the system of passive cooling, and the mass of the whole instrument, in this case, will be no more than 20-30 kg. Besides which superlight spacecraft can be launched on the high-apogee orbit out of the

TABLE 1. The main physical and technical parameters of germanium spectrometer.

Parameter	Value
Energy Range	0.01−10 MeV
Effective Area (E=661 keV)	100 cm^2
Energy Resolution (E=1 MeV)	≈ 0.2%
Field of View	± 90°
Mass	≈ 20 kg
Size	0.2x0.2x0.2 m^3
Power Expenditure	< 20 Wt

Earth magnetosphere, where the homogenous background conditions will be also provided.

Let us consider the advantages of observations of Galactic gamma-ray lines in experiments on-board superlight spacecraft with the use of germanium detector with effective area ≈ 40 cm^2. We will suppose that half of the sky is screened by the spacecraft itself thus the instrumental background is caused only by the cosmic diffuse gamma-quanta falling onto detector from the other "opened" half of the sky. Then by the condition of the constant orientation of the instrument toward the Galactic Centre, the minimal exposure time necessary to detect gamma-ray lines 0.511 eV, 1.8 eV will be ≈ 100 s and ≈ 300 s correspondingly. Taking into account that estimations were made for rather ideal conditions, the real exposure time necessary to detect those lines will be higher. But the order of magnitiude of its value is about a fraction of an hour, thus it is possible not only to detect Galactic gamma-ray lines, but to study its dynamics in this experiment.

It is necessary to note also, that the spectrum of the cosmic gamma-ray background can be also measured precisely in the wide energy range during this experiment. Besides the spectral peculiarities of some temporal phenomena such as cosmic gamma-ray bursts, transients, etc., can also be studied. There are no specific conditions required for orientation and stabilization of a spacecraft.

POLARIZATION MEASUREMENTS

Till the present time, polarization measurements of hard radiation on-board spacecraft were made mainly in the "classical" X-ray range, i.e., less than 20 keV, with the use of Bragg spectrometers and Li or Be detectors. This technique is based on the Thompson scattering of detecting quanta. Of modern projects, Spectrum-Roentgen-Gamma (SRG) is of especial interest. The facility includes a Bragg spectrometer and lithium polarization detector of a large area (1000 cm^2) which should be placed at the focal plane of the soft X-ray concentrator [9].

As the result of previous measurements, polarization of synchrotron radiation from the Crab nebula in the X-ray range at the level 15%, and polarization of X-rays from Cyg X-1 binary at the level 2-5% were discovered [10]. Besides objects like ordinary pulsars, such as the pulsar in the Crab nebula (hard radiation was observed now from about 10 such objects), the X-ray pulsars in binaries and black hole candidates like Cyg X-1 are of particular interest for polarization measurements.

Unlike the ordinary pulsars for which hard radiation is synchrotron, the X-rays from pulsars in X-ray binaries have a thermal origin. They are caused by the hot plasma electrons bremsstrahlung. Regions of a hot plasma are formed near the pulsar magnetic poles as the result of accretion of a matter from the optical companion. The temperature of these regions may be sufficient high, thus the values of parameter kT in the case of two-parametrical spectral approximation are about $\approx 20 - 40$ eV. The nonthermal components caused by the bremsstrahlung of energetic electrons are observed in the spectra of several pulsars. Such electrons can propagate as abeams in the strong pulsar magnetic field. Thus in this case the strong polarization of bremsstrahlung can be expected. Although the probability of plasma thermal radiation polarization is rather low, in some cases it can be observed, if, for example, collective reflection of radiation from the pulsar surface take place.

Thus it seems, that polarization measurements of hard X-rays (energy range 20-100 keV) from pulsars in binaries can give significant information about physical processes in such systems. Polarization measurements are also of particular interest for study of such sources as X-ray binaries - black hole candidates. As a rule, such systems have a rather hard nonthermal spectra, which can be formed as the result of Comptonization of accretion disk infrared radiation. The discovery of polarization of such radiation is of a great meaning for the testing of the models of such objects.

It seems that at the first stage of observations polarization measurements should be realized for very bright sources. Amongst X-ray binaries of the all mentioned above types the typical most bright sources are characterizes by the fluxes about 100 mCrab in the energy range 20-100 keV. For the reasons discussed above the optimal spacecraft for such measurements is superlight carrier with minimal own background.

Let us consider the advantages of polarization experiment with the use of a detector based on a beryllium crystal of 10 cm thickness and ≈ 100 cm^2 effective area. Such relation between thickness and area of scatterer provides the optimal conditions for quanta scattering in crystal with the efficiency 10%. We suggest that during measurements detector constantly oriented toward the observing sources and due to the special collimators field of view (FOV) is rather narrow, 0.1 sr, thus instrumental background is caused only by the cosmic diffuse gamma-quanta falling inside FOV. If we suppose that the background gamma-quanta falling onto detector outside its FOV are absorbed completely, and registration efficiency of scattered in beryllium crystal quanta is about 100%. Then the minimal exposure time necessary to detect a 10% polarized component at the 5σ level is about $\approx 4 \times 10^3$ s; and to detect 1% is about $\approx 4 \times 10^5$ s for the source with flux about 100 mCrab ($\approx 1.4 \times 10^{-1}$ phot/cm^{2}s) in the energy range 20-100 keV.

Hence, this experiment gives some opportunity to measure polarization of Galactic X-ray sources hard radiation for exposure time of the order of days and hours,

TABLE 2. The main physical and technical parameters of polarization detector.

Parameter	Value
Energy Range	20–100 keV
Effective Area of scatterer	100 cm^2
Efficiency of linear-polarized radiation detection	$\approx 10\%$
Field of View	0.1 sr
Mass	≈ 20 kg
Size	0.5x0.5x0.3 m^3
Power Expenditure	< 20 Wt

that could be realized during continuous observations. If the experiment will be continued during several months, then practically all bright X-ray binaries (whose number on the sky is not large, ≈ 20) can be observed. At last, when the exposure time of one source will be several years, it will allows us to search that part of electromagnetic radiation which can be produced as the result of interaction of high-frequency gravitational radiation of neutron star with its magnetic field.

The main difficulties of such observations are connected with the necessity to provide during the long time the constant orientation of the spacecraft with accuracy at least no worse than one degree.

We think that results of these observations will be unique, because until the present time there were not any regular measurements of polarization hard radiation from astrophysical objects as well as its time variations. These new results will allow us to solve some problems of fundamental physics of X-ray binaries and to evaluate the upper limit of intensity of high-frequency gravitational radiation analyzing the presence or absence of some character peculiarities, which will be discussed below.

GRAVITATIONAL RADIATION EFFECTS

Rotating neutron star is also a gravitational wave detector. As is wellknown [11], in neutron stars the photoproduction of gravitons in star matter particles' Coulomb and magnetic dipole fields will take place. The gravitational radiation intensity of neutron star is higher than electromagnetic luminosity of the Sun. The maximum of the neutron star gravitational radiation intensity is at the soft X-ray range, although the intensity at the ultraviolet and optical ranges is also significant. In particular, for wellknown pulsar in the Crab nebula the intensity of gravitational radiation in the ultraviolet and optical ranges is [11] $F = 3 \times 10^{26}$ erg/s, and at the soft X-rays [11] $F = 6 \times 10^{34}$ erg/s. Thus stars are the source of the weak gravitational waves, and resulting gravitational radiation can belong to the any frequency range in dependence of the spectrum of initial photons. The maximum of gravitational radiation intensity will be at the soft gamma-ray range.

Many stars have a field coinciding with the field of the rotating magnetic dipole, which magnetic axis is inclined to the rotation axis. Following from detailed calculations [12 - 14], the result of interaction of high-frequency gravitational wave within a neutron star with its magnetic field, the electromagnetic wave of the same frequency is produced, but the amplitude of this wave will be modulated with the star rotation frequency.

Thus high-frequency electromagnetic wave will come to the observer as the consequence of pulses (analogous to the radiation of pulsars). The shape and frequency of those pulses will depend on the angle β between the vectors of the star magnetic dipole momentum $\vec{m}$ and angular velocity of its rotation ω_0, s well as on the place of the point of observations.

In the common case the pulses of electromagnetic radiation will consist of pulses with frequency ω_0, and $2\omega_0$,, and the amplitude and phase of each of them will depend on the coordinates of the point of observation and angle β. It is necessary to note that this amplitude modulation give only one "range" in which high-frequency electromagnetic wave could be observed. The source of gravitational waves can produce gravitational waves with nonregular amplitudes, thus radio, optical or X-ray, gamma-telescope mounted on-board spacecraft will detect produced electromagnetic wave only, if sufficiently power burst of high-frequency gravitational wave will be in that "range".

Let us consider once more peculiarity of produced radiation: this radiation is polarized linearly, but consists from two parts. The polarization plane of one part is immovable, but of the other part is rotating with the star rotation frequency. Thus the polarization of electromagnetic radiation, produced as the result of interaction of the weak gravitational wave with the field of rotating magnetic dipole is unique. Hence the produced electromagnetic wave can be characterized by the number of typical signs (amplitude modulation, unique polarization state), which allow us to identify its production as the result of interaction of gravitational wave with the field of rotating neutron star.

Now it is rather difficult to indicate the other source except the gravitational-electromagnetic transformations, which will be the cause of that electromagnetic wave peculiarities. But from the other hand, we can not exclude completely the possibility of existence of some other, unknown process in the neutron star magnetosphere, which can result in radiation of electromagnetic waves with above mentioned peculiarities by charged particles. We think that in this situation it is necessary to realize the proposed experiment, to store sufficiently enough massive of experimental data (here it is necessary to take into account, that there are no any experiment, in which polarization of periodic component in neutron star X-rays was measured), and thus the problem of that radiation source nature can be solved as the result of the analysis of obtained data.

Because the maximum of that radiation is at the soft X-ray range, it can observed and studied only by the use of extraterrestrial astronomy techniques. Thus by the realization of astrophysical observations, such as preparing now missions Spectrum UV, and SRG as well as by space research with the use of small satellites, it is necessary to foresee the measurements of polarization states of electromagnetic radiation coming from the neutron stars in all frequency ranges (from radio to hard gamma-rays) and of amplitude modulation of pulses.

In particular, SRG instruments include a concentrator of soft X-rays with quanta energies less than 20 keV. The different detecting instruments are placed at the focal place of this concentrator. The instruments include a polarimeter, which consists of $Li - F$ bars, and has an effective area 0.5 m^2, that allows us to measure polarization of soft X-rays falling onto the instrument. Because the maximum of high-frequency gravitational radiation of neutron stars is just at this range, it seems to include into the SRG scientific program the accumulation of data about polarization of the pulsed component of neutron stars X-rays for the next revealing of the part of this radiation caused by the interaction of gravitational waves with the star magnetic field. Preliminary estimates show that to solve such a problem it is necessary to provide exposure time of about several years. Thus, taking into account the great significance of the gravitational waves detection for fundamental science and technique progress, it seems to justify a separate program of the search of gravitational waves from neutron star, and for this realization it is necessary to foresee the launching of a small satellite which would allow us to observe neutron star radiation during several years. Then the obtained data processing would allow us to search astrophysical sources of gravitational radiation and to estimate the frequency and energy parameters of this radiation, with great meaning for ground observations of these sources.

REFERENCES

1. Ling J.C., Wheaton Wm.A. *Astrophys. J. (Lett.).* **343**, L57 (1989).
2. Sunyaev R.. et al. *Pis'ma v Astron. Zurn.* (Russ.) **17**, 126 (1991).
3. Diehl R. et. al. *Proc. 20th ICRC.* **1**, 144 (1997).
4. *Astrophysika Kosmicheskich Luchey.* Ed. Ginzburg V.L., Moscow: Nauka, 1990 (Russ.).
5. Skibo J.G. et. al. *Astrophys. J.* **397**, 135 (1992).
6. Lamb. R.C. et al. *Nature.* **305**, 37 (1983).
7. Pinto P.A., Woosley S.E. *Nature.* **333**, 534 (1988).
8. Sunyaev R.., erechov .V. *Zemlya i Vselennaya* (Russ.) **2**, 3 (1997).
9. Klapdor-Kleingrothaus J. et al. *Adv. Space Res.* **21**, 347 (1998).
10. oskalenko .I. *tody Vneatmophernoy Astronomii*, Moscow: Nauka, 1984 (Russ.).
11. Papini G., Valluri S.R *Canadian J. Phys.* **53**, 2312 (1975).
12. Denisov V.I. *ZETP* (Russ.). **74**, 401 (1978).
13. Denisov V.I., Yeliseev V.. *P* (Russ.). **65**, 255 (1986).
14. Denisov V.I. In *Experimentalnye testy teorii gravitacii.* Eds. Braginskii V.B. and Denisov V.I., oscow: SU, 1989, p. 102 (Russ.).

Neutron Starquakes

Lucia M. Franco[1,2], Richard I. Epstein[2] and Bennett Link[3,2]

[1] *University of Chicago, 5640 S. Ellis Ave., Chicago, IL 60637*
[2] *Los Alamos National Laboratory, NIS-2 MS D436, Los Alamos, NM 87545*
[3] *Montana State University, PO Box 173840, Bozeman, MT 59717-3840*

Abstract. The Crab and other pulsars suffer sudden and permanent increases in their spin-down rates, suggesting that the torques upon them grow in steps. Torque changes could come about as a consequence of *starquakes* occurring as the star spins down and its rigid crust becomes less oblate. We study the evolution of strain in the crust, the initiation of starquakes, the effects on the magnetic field structure and the observable consequences for neutron star spin down. We find that the stellar crust begins breaking at the rotational equator, forming a fault along which matter flows to reduce the equatorial circumference. Magnetic stresses favor fault lines inclined at an angle to the equator and directed toward the magnetic poles. The resulting asymmetric matter redistribution produces a misalignment of the angular momentum and spin axes. Subsequently, damped precession to a new rotational state increases the angle between rotation and magnetic axes. The change in this angle could increase the external torque, producing a permanent increase in the spin-down rate.

INTRODUCTION: OBSERVATIONAL EVIDENCE FOR STARQUAKES

Most isolated pulsars do not spin down in a regular fashion, but suffer sudden jumps, *glitches*, in their spin rates. Persistent increases (*offsets*) in spin-down rate following glitches have been observed in the Crab pulsar, PSR 0355+54 and PSR 1830-08 (see Table 1). In the Crab pulsar, these permanent *offsets* typically involve fractional changes in the spin down rate of $\sim 10^{-6} - 10^{-4}$ (see Fig. 1). These persistent increases are suggestive of *permanent increases* in the external torque acting on the star. Crust motions, associated with breaking of the rigid crust as the star slows down, could change the orientation of the magnetic moment with respect to the spin axis and hence change the external torque. Here we describe how a neutron star breaks and relaxes its structure as it spins down, and consider the consequences for torque evolution.

TABLE 1. Observed persistent shifts in spin-down rate: The magnitudes of observed persistent shifts for the Crab and two other pulsars. All observed persistent shifts are of the same sign, and correspond to *increases* in the spin-down rate. A particularly dramatic *offset* followed the Crab glitch of 1989. The magnitude of the offset in PSR 0355+54 is uncertain since the pulsar has not had time to completely recover from the glitch. (Data from [4] and [6]).

PSR	age (yr)	glitch year	$\Delta\Omega/\Omega$	permanent offset $\Delta\dot\Omega/|\dot\Omega|$
Crab	10^3	1969	4×10^{-9}	-4×10^{-6}
		1975	4×10^{-8}	-2×10^{-4}
		1986	4×10^{-9}	-2×10^{-5}
		1989	9×10^{-8}	-4×10^{-4}
0355+54	6×10^5	1986	4×10^{-6}	-3×10^{-3}?
1830-08	2×10^5	1990	2×10^{-6}	-8×10^{-4}

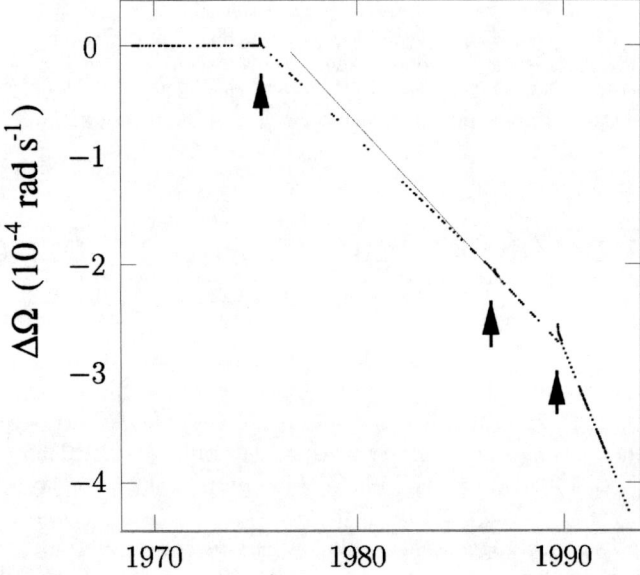

FIGURE 1. Twenty-five years of spin history of the Crab pulsar showing apparent torque increases (adapted from [4]). Shown are spin rate residuals relative to a model for data prior to the first glitch. At each glitch, indicated by arrows, the star acquired a permanent increase in spin-down rate.

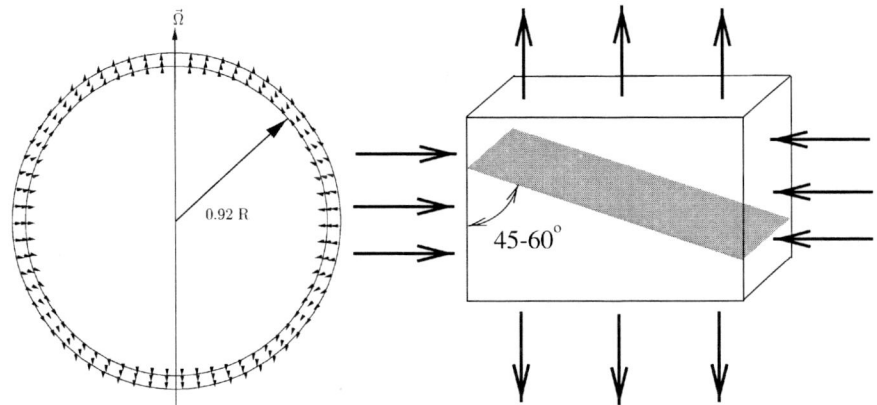

FIGURE 2. Left: Displacement field in a spinning-down neutron star with a crust thickness of 8% of the total stellar radius. A cross-section through the center of the star is shown. The equatorial diameter decreases, while the polar diameter increases. **Right:** A block of matter on the equator is under relative compression and tension. The block shears along a plane when the critical strain is reached. Shearing along a complementary plane flipped over with respect to the plane shown is equally likely for isotropic material.

HOW DO STARQUAKES OCCUR?

We model the star as a two-component homogeneous spheroid: a brittle crust of uniform density afloat on an incompressible liquid core. In equilibrium, the star has a spheroidal shape with an equatorial bulge. As the neutron star spins down, the fluid interior becomes more spherical, while strain develops in the rigid crust. The crust breaks once the strain reaches the elastic limit for the crustal material. The distribution of strain determines the geometry of the starquake. We find that the matter displaces as shown in Fig. 2 (left panel).

A calculation of the strain field shows that, for a crust thickness $< 8\%$ of the stellar radius, the strain is largest at the equator, where a fault will begin forming once the crust breaks [7]. There, an element of matter receives compressive stress until it shears[1] (Fig. 2, right panel). The break begins somewhere on the equator, along fault f or f' of Fig. 3.

---

[1] Known materials exhibit ductile rather than brittle behavior at pressures comparable to their shear moduli [2]. However, deep-focus earthquakes are known to originate from regions of very high pressure [8]. These faults are thought to be facilitated by densification phase changes; small regions of higher density nucleate in the stressed material, and act as a lubricant for shearing motion. Analogous processes might occur in the high-pressure material of the neutron star.

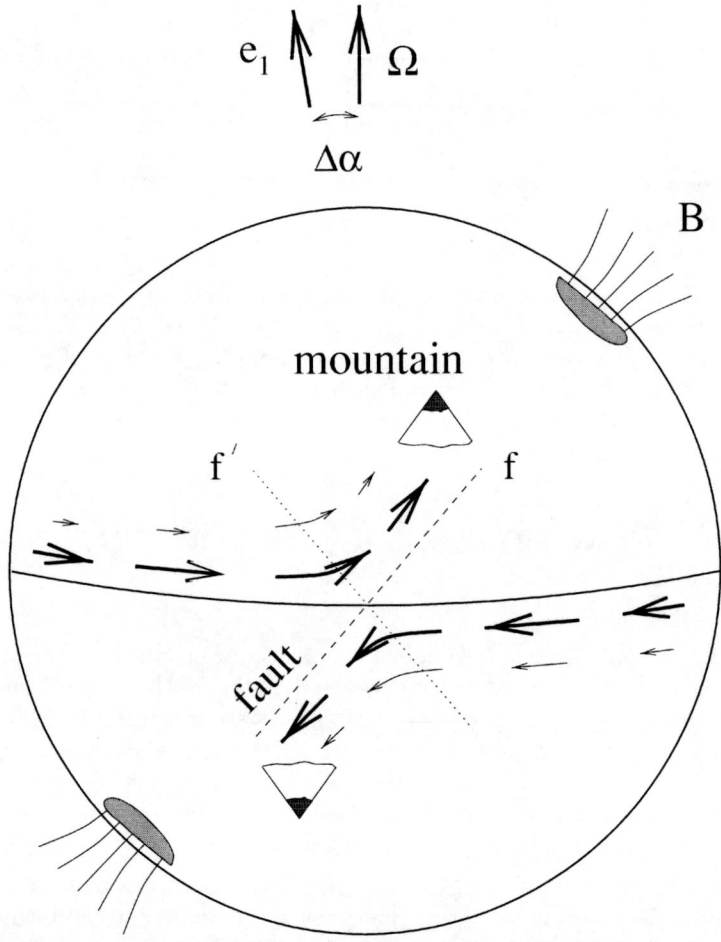

FIGURE 3. Fault propagation in the presence of a strong magnetic field occurs preferentially along f, creating "mountains" (indicated by the snow-capped peaks) and shifting the largest principal axis of inertia to a new direction e_1 (fixed in the star). Breaking in a similar fashion can occur on the back of the star.

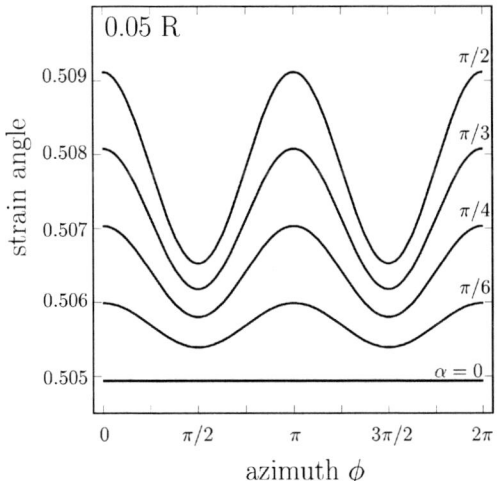

FIGURE 4. The strain angle of surface equatorial material in the presence of a magnetic field (arbitrary units) for a crust of thickness 5% of the total stellar radius. The curves correspond to different values of the angle α between the magnetic and rotation axes. The magnetic poles are at azimuthal angles $\pi/2$ and $3\pi/2$.

MAGNETIC EFFECTS BREAK THE SYMMETRY

The discussion so far considers rotational and gravitational effects only and takes no account of the magnetic field of the neutron star. The matter motions induced by the spin down distort the magnetic field anchored in the crust, generating magnetic stresses that affect the development of strain. Furthermore, the rotational and gravitational forces possess azimuthal symmetry which will be broken when we consider the magnetic axis inclined at an angle α to the rotation axis.

For azimuthal symmetry, there is no preferred location on the equator for the break to begin. Fig. 4 shows the strain angle including both spin-down and magnetic stresses. The magnetic field inhibits strain close to the magnetic poles, making starquakes most likely to originate at the two points on the equator farthest from the magnetic poles. From there, matter will move to higher latitudes allowing the star's equatorial circumference to decrease.

THE BRAKING TORQUE IS INCREASED

At the equator the starquake is equally likely to begin along the faults f and f'. The magnetic field, however, inhibits motion across field lines, and fault f' is less favored than f. As material moves along fault f, "mountains" ($< 10\ \mu$m high) form at higher latitudes, breaking the axial symmetry of the star's mass distribution (see Fig. 3). The mass motions shift the principal axis of inertia e_1 of the star by an angle $\Delta\alpha$ *away* from the angular momentum vector. The star's rotation and angular momentum vectors are now misaligned and the star will precess [9]. The damping of the precession will restore alignment of the principal axis of inertia and the angular momentum vector, increasing the angle α between the rotation and magnetic axes to $\alpha + \Delta\alpha$. In many models of pulsar spin down (e.g., the vacuum dipole model), an increase in α produces a *permanent* increase in the spin-down torque.

We have estimated that the release of the spin-down strain that develops between glitches in the Crab is adequate to produce $\Delta\dot{\Omega}/|\dot{\Omega}| \sim 10^{-6} - 10^{-4}$ [10], as observed. There is a number of observational signatures one would expect from a cracking of the neutron star crust such as we have described here. The damped precession caused by the asymetric redistribution of matter is of order a few seconds [7] and we don't expect it would be observable except in the fortunate case in which a glitch occurred during observations so that enough sampling can unveil precession on the order of minutes [5]. The reorientation of the magnetic axis translates into a reorientation of the radio emission beam with respect to the observer. This may produce a change in the pulse profile of the pulsar and an associated change in the flux of order $\sim 1\%$ [3] which may be measurable both in x-ray and optical bands. Crustal quakes are expected to excite Alfven and fast magnetosonic waves in the magnetosphere of the neutron star. However, estimates of the energy that will thus be converted to radiation and the details of the conversion mechanism are too uncertain [1].

SUMMARY

The Crab and other pulsars show evidence for external torque increases coincident with glitches. We suggest that these torque increases are a consequence of *starquakes* driven by the star's spin down. The calculated geometry of starquakes in a spinning-down neutron star is shown in Fig. 3. The resulting formation of "mountains" could produce an increase in the external torque of the magnitude implied by the data.

ACKNOWLEDGEMENTS

This work was performed under the auspices of the U.S. Department of Energy, and was supported in part by NASA EPSCoR Grant #291748, by IGPP at LANL and by the Center for Astrophysical Thermonuclear Flashes at the University of Chicago.

REFERENCES

1. Blaes, O., Blandford, R., Goldreich, P. & Madau, P., *ApJ*, **343**, 839 (1989).
2. Duba, A. G., *et al.*, *The Brittle-Ductile Transition in Rocks: the Heard volume*, AGU: Washington, 1990.
3. Link, B., and Epstein, R. I., *ApJ*, **478**, L91 (1997).
4. Lyne, A. G., Pritchard, R. S., & Smith, F. G., *MNRAS*, **265**, 1003 (1993).
5. McCulloch, P. M., Hamilton, P. A., McConnel, D., & King, E. A., Nature, **346**, 822 (1990).
6. Shemar, S. L. & Lyne, A. G., *MNRAS* **282**, 677 (1996).
7. Franco, Lucia M., Link, B. & Epstein, R. I., submitted to *ApJ* (1999).
8. Green, H. W. II & Houston, H., *Ann. Rev. Earth Planet Sci.* **23**, 169 (1995).
9. Shaham, J., *ApJ* **214**, 251 (1977).
10. Link, B., Franco, Lucia M. & Epstein, R. I., *ApJ* **508**, 838 (1998).

The ROTSE Detection of Early Optical Light from GRB 990123

Galen Gisler*

On behalf of the ROTSE collaboration:

Carl W. Akerlof[†], Richard J. Balsano*, Jeffrey J. Bloch*, Donald E. Casperson*, Sandra J. Fletcher*, Galen R. Gisler*, Jack G. Hills*, Robert L. Kehoe[†], Brian C. Lee[†], Stuart L. Marshall[‡], Timothy A. McKay[†], Richard S. Miller*, William C. Priedhorsky*, John J. Szymanski*, and James A. Wren*

*Los Alamos National Laboratory[1]
[†] University of Michigan
[‡] Lawrence Livermore National Laboratory

It is perhaps odd to present a paper on a ground-based observational astronomy project at a workshop on small satellites, but the Robotic Optical Transient Search Experiment (ROTSE) is indeed a small project (if not a small satellite), and we have recently obtained a result of high interest to the small-satellite community. This result was the detection of very early time light from a gamma-ray burst, at a brightness accessible to very small telescopes. We argue that this result should influence the design of future gamma-ray burst missions, both on the ground and in space. The details of our observation of GRB 990123 have been published, subsequent to this workshop, in the journal *Nature* (Akerlof et al. 1999), and I therefore present in this paper an anecdotal review of gamma-ray bursts, our system, this particular event, and subsequent developments in our project.

Gamma-Ray Bursts (GRBs) were discovered in the early 1970's by the Los Alamos series of Vela satellites, which had been looking for violations of the limited nuclear test-ban treaty. Many satellite missions since those early days have studied them, but they remain among the most puzzling phenomena in the cosmos. By far the most prolific source of data on GRBs has come from the Compton Gamma-Ray Observatory, primarily the Burst and Transient Source Experiment (BATSE) on board that satellite. By this time, many thousands of bursts have been observed, and new ones are discovered at the rate of about one per day. New generations of satellites, including HETE, SWIFT, BALLERINA (all discussed at this meeting) will be addressing these phenomena in the future.

---

[1] email: gisler@lanl.gov

Since the beginning of the study of GRBs, attempts have been made to observe these events at other wavelengths, principally using optical telescopes on the ground. Archive searches and photographic patrols yielded no counterparts, and all attempts to perform simultaneous observations were hampered by the poor positional information obtained by the gamma-ray observations, by the short duration of the events, and by the difficulty in disseminating alerts quickly.

Because of the lack of observations at other wavelengths, understanding of these phenomena developed only slowly. Phenomenologically, GRBs are intense, brief, localized emissions of radiation at energies from 10s of keV through 10s of MeV, with isolated instances of reported detections at 10s of GeV and even TeV energies. Their durations range from 10s of milliseconds to hundreds of seconds, with a hint of a division into two populations separating around a few seconds. When they occur, they are by far the brightest objects in the sky at their wavelengths. Because of their high brightness, it has been possible to study their light curves and gamma-ray spectra in great detail. They have an enormous diversity in their light curves, from simple isolated pulses to complex and multiple structures; sometimes their are precursors, sometimes aftershocks, and the spectral energy distributions have been observed to change in a variety of ways during a burst. Unfortunately, though the light curves are well characterized, they don't provide much information to aid understanding. They may not be a homogeneous population, yet they differ profoundly from all other known astrophysical phenomena in being transient, violent, and nonrepetitive. The conservative approach is to treat them as a relatively homogeneous population until we know better.

Population studies of GRBs have led to the conclusion that they are very distant, probably cosmological. Firstly, they are isotropic to a degree shared by no population of relatively nearby known sources, and secondly, the number-brightness distribution shows a deficit of faint sources as might be produced by cosmic evolution.

Theoretical understanding of GRBs has been slow, though there have been hundreds of papers devoted to the subject, many developing hypotheses that now seem rather quaint. The gradual acceptance of a cosmological distance scale for GRBs has limited the playing field to those theories that can provide the prodigious energies that are required. Collisions of neutron stars with each other, or with black holes, or processes leading to the rapid formation of a black hole, such as special kinds of supernovae, or hypernovae, seem to be among the leading candidates, though the processes by which much of this energy is liberated in the form of gamma-rays remains tentative and speculative.

Above all, we've needed more information on these objects, particularly at different wavelengths, but that has been very hard to come by. Without good positions for GRBs, they cannot be identified with other known types of sources. But it is difficult to get good positions from the gamma-rays alone. Our detectors are not sufficiently discriminatory in direction. Our main tool for getting directions has been the use of timing and intensity ratios among different detectors, or elements of a single detectors. The BATSE detectors provide error circles of degrees in this way, but only with interplanetary baselines is this technique good enough to provide error circles of arc minutes, enabling identifications. Until 1997, the Interplanetary Network (IPN), involving several satellites at different locations in the solar system, was the only source of good GRB positions. However, IPN positions are never available until hours (if not days or weeks) after the burst occurs, prohibiting simultaneous observations.

The lack of good rapid positions led to the development of wide-field rapid response optical telescope experiments and a network of rapid alert dissemination via the internet, first known as Bacodine (BATSE Coordinate Distribution Network) and then as GCN (Gamma-ray Coordinates Network). Among the optical experiments that were deployed and subscribing to GCN were the Explosive Transient Camera (ETC, Krimm et al 1996), the Gamma-Ray Optical Counterpart Search Experiment (GROCSE, Lee et al 1997), the Livermore Optical Transient Imaging System (LOTIS, Park et al 1997), and our own system, ROTSE. Collectively, these wide-field experiments ran for a dozen years without seeing any optical counterparts to GRBs.

The strategy of the wide field search experiments was essentially to have a small automated telescope in a constant state of readiness (generally performing background observations); it receives coordinates of a GRB very quickly from GCN, and if the object is in an accessible part of the sky, it slews to the indicated position and performs a sequence of exposures, possibly tiling to look at an even wider area of sky. Afterwards, the entire exposed field is searched for a transient or variable source. This approach was mostly unsuccessful, though useful limits were placed on a few GRB optical counterparts.

Our ROTSE projected consisted of two separate systems, ROTSE I, a 2x2 array of Canon telephoto lenses (f/1.8, 200mm focal length) with large format CCD imagers (Thomson 14 m 2048x2048 pixels), and ROTSE II, a 0.45m, f/1.9 telescope. ROTSE I has a combined field of view of 16.5 degrees, well suited to covering a large fraction of the BATSE error circle. It has been running in automated mode since March 1998, and has so far accumulated more than 2 Terabytes of data in its background mode of continuously patrolling the overhead night sky. It has responded to a few dozen alerts, with an average response time of about 10 seconds from the BATSE trigger time (including an average of 3 seconds slewing time). ROTSE II has a field of view of just under 2 degrees, and is not yet working automatically.

The major breakthrough in GRB studies occurred in 1997 with the first detection of an optical afterglow of a gamma-ray burst, but this was not accomplished by any of the wide-field searches. Other teams of optical and radio observers had subscribed to GCN with the hope of seeing afterglows of GRBs with conventional narrow-field large-aperture telescopes. They succeeded in discovering the first optical counterpart when a source of accurate positions for GRBs became available. This source was the Italian-Dutch Beppo-SAX satellite, which had a GRB monitor plus a wide-field X-ray camera. The rough positions given by the monitor were sufficient to point the X-ray camera, which in turn gave positions to a few minutes

FIGURE 1. ROTSE I atop its enclosure, pointed at zenith

of arc. Because this satellite was not in constant communication with the ground, and because the repointing took some time, the accurate positions were not available until several hours after the burst occurred. But this was significantly more rapid than the IPN positions, which were of similar accuracy.

The first two optical counterparts found, GRB 970207 and GRB 970508, by van Paradijs, Kulkarni, Frail, and others, using the Beppo-Sax positions, proved a sensational boost to this field. We finally had direct confirmation, via the redshift of 970508, that at least some GRBs (and probably all) were at cosmological distances. The study of afterglows became an industry in itself, with more observers signing up to perform counterpart observations with their large telescopes. A dozen optical afterglows were discovered over the next two years, with some radio afterglows as well, and redshifts were measured for roughly half. Several common features were noted, including power-law decays, indicating deceleration by an external medium. All were in host galaxies, though not centrally located, and those with measured redshifts were all at distances greater than $z \sim 0.8$ (except for one, GRB 980425, which was associated with a supernova in a relatively nearby galaxy). One burst, GRB 971214, holds the record for the most distant, at $z = 3.4$. With actual measurements of distances, and radiation in new wavelength regimes, came new impetus for theoretical developments. The energy requirements were now accepted to be greatly in excess of supernova energies, and theoretical models now looked toward black hole formation scenarios. Additionally, the fireball models became more sophisticated and interesting, involving external shocks and internal shocks

FIGURE 2. ROTSE I (left) and ROTSE II with clamshells partially open.

as well as a central cataclysmic source. Comparisons of time structures in the optical and radio led to new insights, and radio scintillation observations gave the first measurements of the expansion of the fireball.

Unfortunately, it also seemed clear that large telescopes were indeed needed to play in this game, because the afterglows themselves were very faint. In particular, 970508 had apparently been caught on the rise, peaking at 19th magnitude. Things were looking very bad for wide-field early-time searches for gamma-ray optical counterparts.

In ROTSE, we worked on placing early limits on some bursts, and looked toward changing our observing strategies to achieve the depth that appeared necessary to find the late-time afterglows. This involved longer exposures, less tiling, sacrificing coverage of the extended error field, etc. A delicate problem we faced was: what if we found a very bright and rapidly fading transient, at early times, somewhere in the 256 square degrees covered by ROTSE I. Would anyone believe us? Though we were looking in a unique region of parameter space, we were ourselves doubtful.

Then Nature presented us with a gift. On 1999 January 23, 09:47 UT (= 02:47 MST) BATSE and Beppo-SAX both triggered on a strong GRB. Four seconds later, the GCN signal arrived at ROTSE I, which terminated its sky patrol exposure and began steering toward the early BATSE coordinates (which were 9 degrees away from the subsequent localization). By ten seconds after the burst trigger, ROTSE I had begun its first 5-second exposure on the target. Unfortunately, this exposure, and all subsequent odd-numbered exposures, were lost due to software errors. The second exposure, and the first one saved, was begun 22 seconds after the BATSE trigger, recording a transient stellar object at 12th magnitude. The next saved exposure, at 47 seconds after the trigger, recorded the same transient at 9th magnitude. Since this burst was a long burst, this and two other optical frames were obtained while the gamma-rays were still being received by BATSE. Jim Wren, who was covering shift that Saturday morning, was awakened by his pager, which reported that ROTSE I was responding to a BATSE alert. He verified that the system was working, and went back to sleep. By 45 minutes after the initial trigger, the ROTSE I alert response was finished, and the system resumed the interrupted sky patrol. Our transient detection was in our data, but as yet undiscovered.

Four hours after the grigger, Beppo-SAX reported a position for the burst that was correct to 5 minutes of arc (Piro et al *GCN Circ. No. 199*). Immediately, Odewahn, Bloom and Kulkarni used a 60-inch telescope to point to that position, and identified an optical transient at 18th magnitude, which they promptly reported (Odewahn et al *GCN Circ. No. 201*). Eight hours after the trigger, Carl Akerlof and Tim McKay of our collaboration used the Beppo-SAX position to search our data, and they found the transient at the same position reported by Odewahn et al, but much brighter. This was likewise promptly reported (Akerlof and McKay *GCN Circ. No. 205*). The report of the redshift, z = 1.6 (Hjorth et al, *GCN Circ. No. 219*), came a day later, and proved this burst to be the most luminous optical flash ever seen, and the most violent by far.

The discovery of simultaneous optical emission in a gamma-ray burst at such a brightness was not entirely unexpected. Some theorists had been encouraging early-time searches for several years (e.g. Meszaros and Rees 1997). Nevertheless, it is fair to say that this event has renewed efforts to cover the time domain in astrophysics to a much greater extent than was previously conceived.

Unfortunately, the discovery of early bright transients is still not favored by the present suite of detectors. ROTSE I needed both the early BATSE position to point the telescope, and the subsequent Beppo-SAX localization to find the 990123 transient in the wide field. Because the early BATSE position was off by 9 degrees, we very nearly missed it! Future satellites will give more accurate early positions, but we must stress the importance of getting those positions to the ground as rapidly as possible. Some proposed future satellites will have co-mounted optical telescopes, but because of the necessity of conserving angular momentum while slewing the telescope, rapid response from such a system is very difficult to achieve. Ground-based instruments can slew to a position much more rapidly, easily compensating for the data processing and transmission time. In addition, ground-based instruments are much cheaper and easier to maintain that optical telescopes in space.

We are presently attempting to expand our system to include several additional

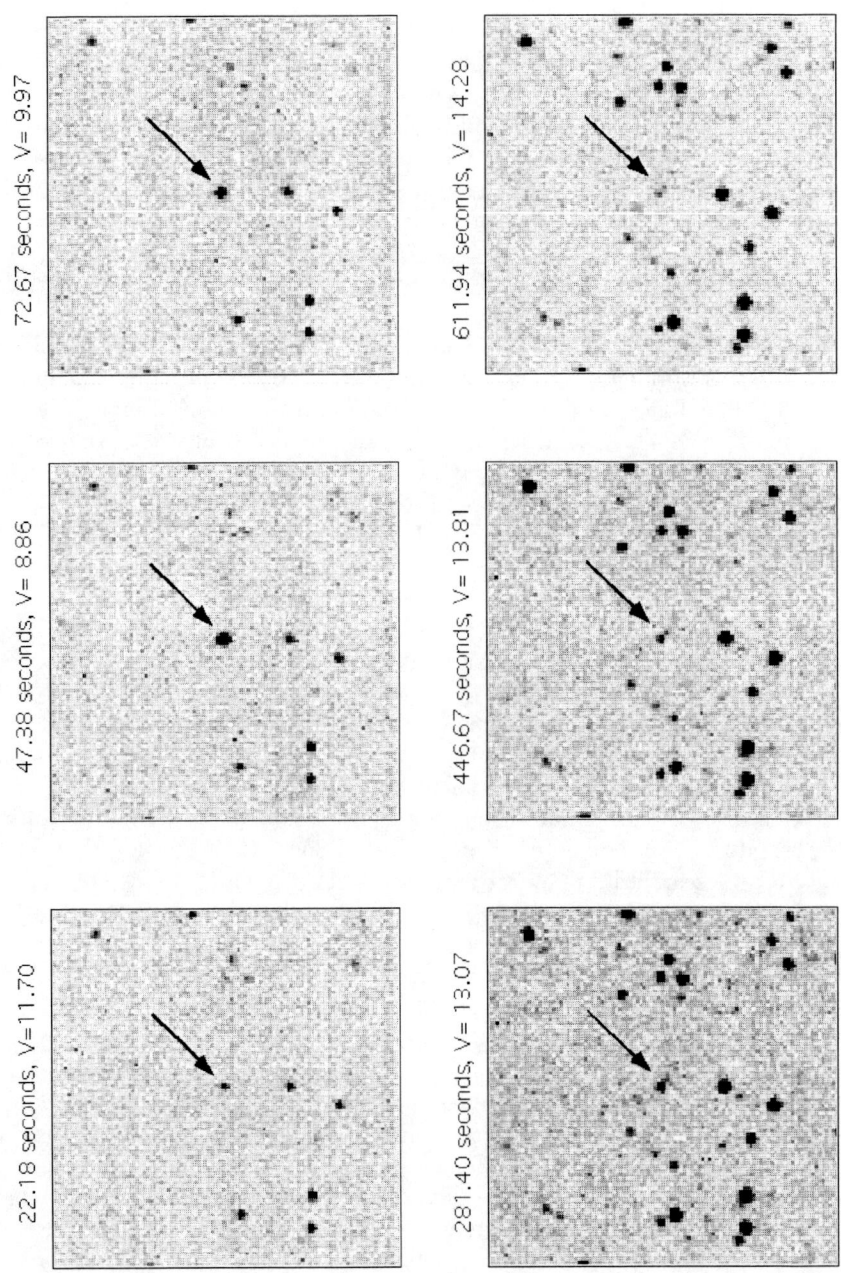

FIGURE 3. The six ROTSE frames for the transient associated with GRB 990123

sites around the world, widely separated in longitude and latitude, so that we may be in a very good position to respond to alerts from the future generation of satellites (HETE II, SWIFT, GLAST).

REFERENCES

1. C. Akerlof, R. Balsano, S. Barthelmy, J. Bloch, P. Butterworth, D. Casperson, T. Cline, S. Fletcher, F. Frontera, G. Gisler, J. Heise, J. Hills, R. Kehoe, B. Lee, S. Marshall, T. Mckay, R. Miller, L. Piro, W. Priedhorsky, J. Szymanski & J. Wren, "Observation of contemporaneous optical radiation from a gamma-ray burst", *Nature* **398**, 400 (1999).
2. C.W. Akerlof & T.A. McKay, *GCN Circ. No.* **205** (1999).
3. J. Hjorth et al, *GCN Circ. No.* **219** (1999).
4. H.A. Krimm, R.K. Vanderspek, & G.R. Ricker, "Searches for optical counterparts of BATSE gamma-ray bursts with the Explosive Transient Camera", *Astron. Astrophys. Suppl.* **120**, 251 (1996).
5. B. Lee et al, "Results from Gamma-Ray Optical Counterpart Search Experiment, a real-time search for gamma-ray burst optical counterparts", *Ap.J.* **482**, L125 (1997).
6. S.C. Odewahn et al, *GCN Circ. No.* **201** (1999).
7. P. Meszaros & M. Rees, "Optical and long-wavelength afterglow from gamma-ray bursts", *Ap. J.* **476**, 232 (1997).
8. H.S. Park et al, "New constraints on simultaneous optical emission from gamma-ray bursts measured by the Livermore Optical Transient Imaging System experiment", *Ap. J.* **490**, L21 (1997).
9. L. Piro et al, *GCN Circ. No.* **199** (1999).

The Intergalactic Medium

Richard C. Henry

Center for Astrophysical Sciences
The Henry A. Rowland Department of Physics and Astronomy
The Johns Hopkins University
Baltimore, Maryland 21218-2686
henry@jhu.edu

Abstract. It is remarkable that a small space satellite mission can be created that has, potentially, the capability of detection of the dark matter of the universe, and in particular, detection of the intergalactic medium. I describe the approach for such a sample mission, and I also briefly comment on, and *illustrate*, black holes; black holes represent another candidate for the "missing" baryonic dark matter in the Universe.

INTRODUCTION

I ask the question, what role can small missions (say, NASA SMEX and MIDEX missions) play in the search for the intergalactic medium and the search for the dark matter of the Universe. I am the Principal Investigator on HUBE (which currently stands for *"Hot Universe Background Explorer."*) HUBE is an example of a mission that one could hope might lead to the direct detection of the intergalactic medium. The high point (so far) in my attempt to implement HUBE, was the acceptance of HUBE by NASA in 1996 as the MIDEX alternate to MAP. HUBE will be submitted once again in the next SMEX round, and of course I very much hope for actual implementation.

THE HUBE MISSION

It is widely agreed that even if the Universe is open, with a small non-zero cosmological constant, there is a very large amount of dark matter in the Universe. As much as 90% of the baryonic dark matter has yet to be detected, and the non-baryonic dark matter, which could make up as much as 90% of the total matter in the Universe, is entirely undetected. This is a very fat target indeed, but one that seems to be very difficult to hit.

While HUBE has many astrophysical goals beyond the hope of detection of the intergalactic medium (please see my paper on the interstellar medium at the present workshop), the search for the dark matter of the Universe is my core reason for the creation of HUBE. The HUBE mission is aimed at spectroscopy of the diffuse ultraviolet background radiation. The existing such observations are collected in Figure 1, which I have adapted and improved from Henry & Murthy (1994). I have kept the labelling of the observations the same as in Henry and Murthy (1994), so that references for the individual points can easily be found by consulting that publication. The main improvement in the figure (other than reversing it to have energy increase to the right) is the inclusion of the new and much lower Voyager (V) upper limit of Murthy et al. (1999) short of 1000 Å.

The reversal of the energy scale allows more convenient comparison of the detailed observations with those in the visible and in the soft X-ray part of the spectrum, which are laid out in Figure 2. Figure 2 also includes the entire regime of operation of HUBE, in spectral terms. The three solid filled points in Figure 2 are those of Bernstein (1998), representing the optical background radiation. The narrow line below her three points are her integration of the galaxies in the same field of view

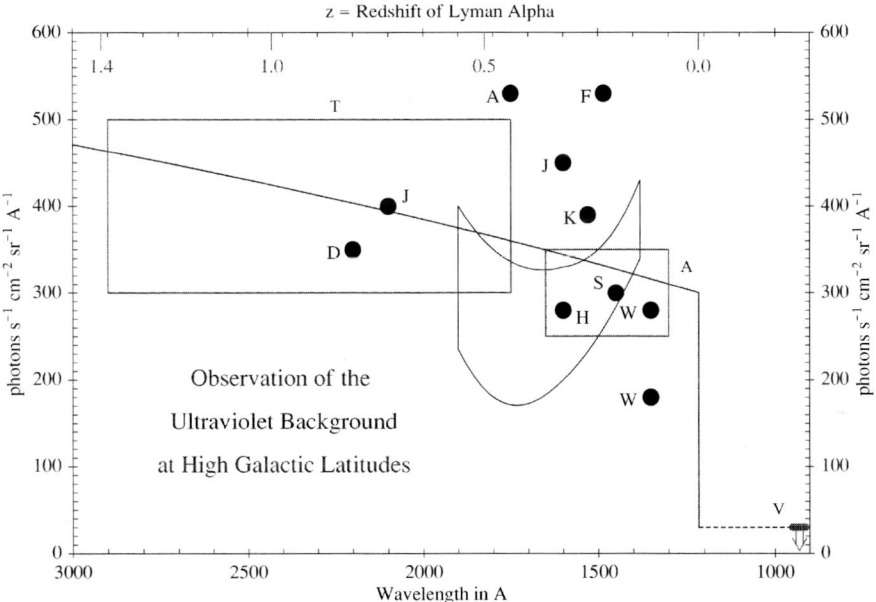

FIGURE 1. Existing Observations of the Diffuse Cosmic Background Radiation. References to individual observations may be found in Henry and Murthy (1994). The most interesting data point is the Voyager upper limit of Murthy et al. (1999), marked with a V. The thin solid line shows a model of the emission expected from clumped clouds of intergalactic gas that expand as the Universe itself expands. The solid points are all actual detections of diffuse radiation, and the spread in values reflects real point-to-point variations in the brightness of the background radiation. In contrast, the Voyager observation is simply an upper limit.

(HDF). A significant diffuse optical background is seen, that is of unknown origin. That this background is truly diffuse has been demonstrated by Vogeley (1998).

Of course the big excitement from the point of view of cosmology, is the large jump in the diffuse background that occurs at (or near) Lyman α. On the face of it, a diffuse background that shows this behavior is begging to be interpreted as being made up of redshifted Lyman α radiation from the missing baryonic dark matter, which would in these circumstances be made up of intergalactic clouds of the type that are discussed by R. Cen at this workshop.

The problem with this interpretation is that, while it is easy to obtain the observed intensity of radiation by clumping the gas to a reasonable degree, an additional result of the clumping is that the recombination time becomes shorter than a Hubble time, and some unknown source of ionization is required to keep the in-

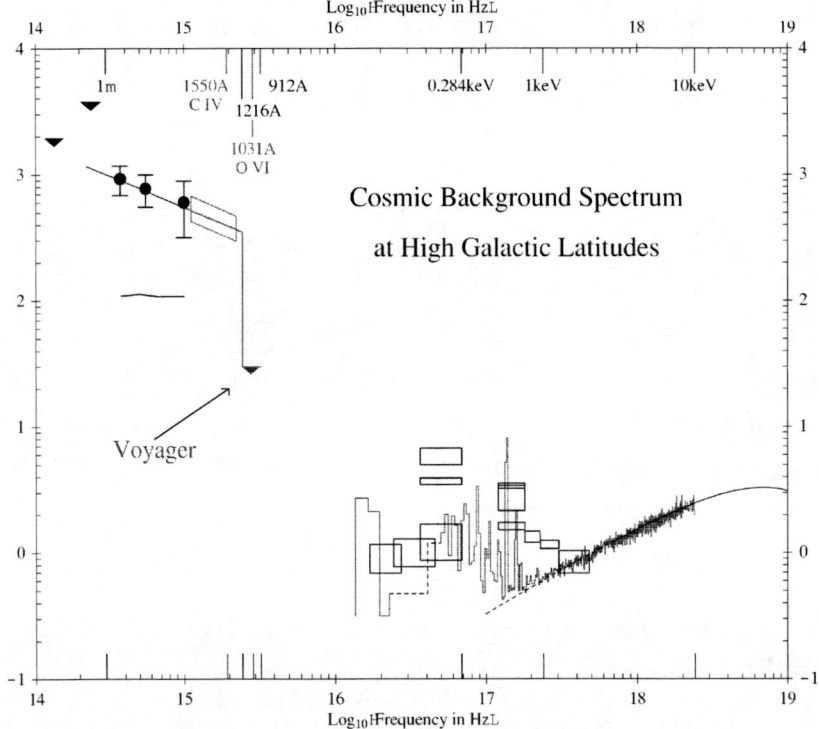

FIGURE 2. The Diffuse Background, from the Visible to X-ray Energies. In this figure, the Voyager and other ultraviolet observations from Figure 1 are shown adjacent to the visible background radiation observations of Bernstein (solid dots). References to other observations in this figure may be found in Henry (1999). The detailed X-ray spectrum is a simulation, due to David Burrows, of what CUBIC (Burrows 1996) could accomplish, if implemented as part of the HUBE project.

tergalactic medium transparent as it is observed to be. I have suggested (Henry 1999) that the putative neutrinos of Sciama (1997) would do the job, but of course I am not averse to any source: *but* the ionizing radiation, *I do need!*

The observations that are given in Figures 1 and 2 were made at high galactic latitudes. The environment (in the ultraviolet) for such observations at high *southern* galactic latitudes is shown in Figure 3. The shading is integrated ultraviolet (1565 Å) radiation from stars. The coordinate system is ecliptic, which is the coordinates of the "worst noise" from the point of view of the *optical* background. Note that the coordinate labels refer to ecliptic coordinates, not to the superimposed galactic coordinates. The interference by direct starlight is not serious at high galactic latitudes, the stars being few and faint. However, the potential for high-galactic-latitude dust back-scattering of ultraviolet starlight is very real, and provides the only realistic alternative to my cosmological interpretation of the observations (assuming that the observations are correct!). That alternate interpretation is, of course, that we are seeing starlight back-scattered from dust.

There are two answers to this, the first being simply that it would be remarkable indeed if the scattering properties of interstellar dust were such that a break in the magnitude of the scattered signal as large as that shown in Figures 1 and 2 should occur at all, much less that it should fortuitously occur near the wavelength

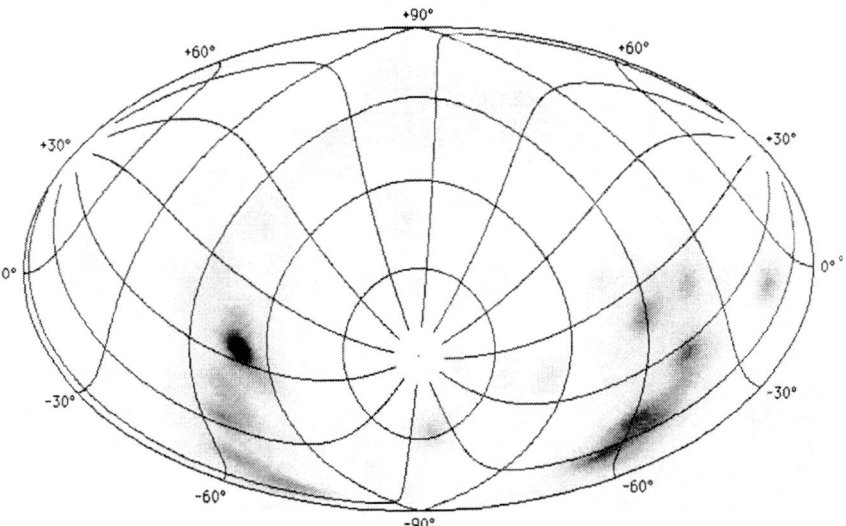

FIGURE 3. The Celestial Sphere in Ecliptic Coordinates (Galactic Coordinates superimposed). The shaded background is the integrated light of stars in the ultraviolet. Zodiacal light is symmetric with respect to ecliptic coordinates, but does not show as much variation with ecliptic latitude in the ultraviolet as it does in the visible (Murthy et al. 1990)

of Lyman α. The second argument is perhaps better, and that is, that Murthy et al. (1994) have observed the Coalsack Nebula using the Voyager ultraviolet spectrometers (both Voyager 1 and Voyager 2), and do *not* see any such break in the spectrum of what is clearly dust-scattered starlight.

A sample of diffuse ultraviolet background radiation is given in Figure 4. This is the spectrum of a high-latitude target observed with the UVX experiment on the Space Shuttle (Murthy et al. 1989, 1990). What is shown in a scan over a few degrees, spectrally resolved from Lyman α (1216 Å) to 3000 Å. Even though a Calcium Fluoride filter is present to attenuate Lyman α, a strong solar-system Lyman α line is seen. This gives an accurate impression of how difficult it is to observe *shortward* of Lyman α, where that emission must be admitted to the spectrometer.

Toward the end of the scan shown in Figure 4 (upper part of the figure), strong terrestrial emission is seen. Throughout the scan, weak terrestrial OI emission at 1304 Å and 1306 Å is also seen. A few horizontal lines are the spectra of stars passing through the field of view. Vertical bands of emission in the right half of the figure are zodiacal light, which declines sharply in intensity in the HUBE spectral region, that is to say, below 2000 Å.

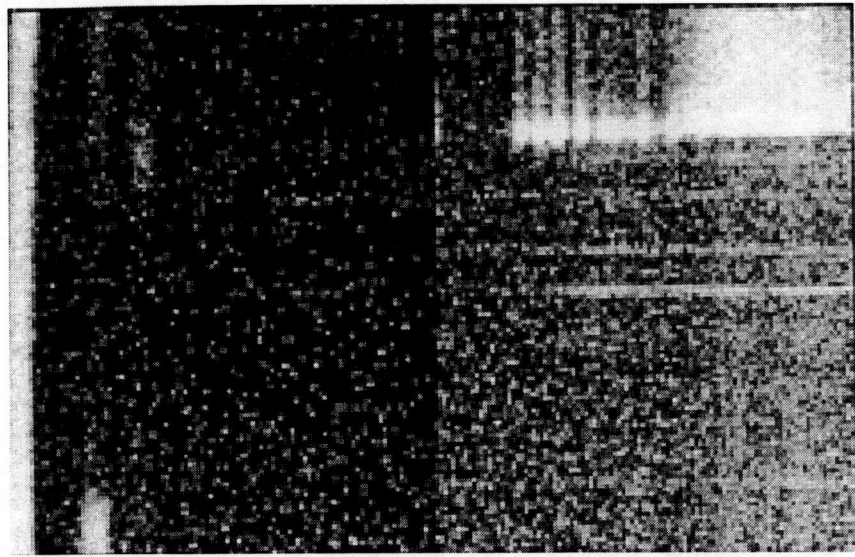

FIGURE 4. An observation of the diffuse ultraviolet background. Various terrestrial features are describe in the text. The sharp break in the background that occurs about half way across the figure, is due to different instrumental backgrounds in the two scanning spectrometers employed aboard the Space Shuttle in making the observations. Control of instrumental background is a key priority of the HUBE mission.

To observationally explore a baryonic intergalactic medium consisting of hot gas clouds in the temperature range $10^5 - 10^6 K$ requires spectrometers spanning the ultraviolet and soft X-ray spectral regions, and that is precisely what HUBE offers (see the X-ray HUBE simulation in Figure 2). What could HUBE teach us about background radiation and the intergalactic medium? If, in the ultraviolet, as is strongly suspected (e.g., Witt and Petersohn 1994), there is an extragalactic component to the background radiation, we would, with HUBE-FUVS continuum measurements (and guided by HUBE-LαS and HUBE-EUVS [these acronyms are identified, and the corresponding instruments characterized, in my paper on the interstellar medium in the present volume] step-size measurements), create a three-dimensional map of emission from the nearby ionized intergalactic medium. Simultaneously, with HUBE-CUBIC, we would make the definitive measurements of the Cosmic X-ray Background Radiation (CXRB) spectrum, in a search to understand its origin. All previous CXRB mappings (Wisconsin rockets, SAS-3, HEAO-1, and ROSAT) used proportional counters, which have very poor energy resolution. In only six months' exposure, HUBE-CUBIC would measure the spectrum of the CXRB with larger total exposure than HEAO-1 but with an eight-fold improvement in spectral resolution. (Sensitivity to diffuse emission depends on the instrument *étendue*, AΩ: HUBE-CUBIC is significantly more sensitive than any previous mission other than HEAO-1.) Our excellent spectral resolution combined with high sensitivity would be vital for understanding the origin of the CXRB: it is believed (Ueda et al. 1998) that most of the flux in the 1-2 keV band comes from weak sources at high z, primarily AGN's. Iron K-line emission from AGN sources should produce structure in the CRXB spectrum strongly indicative of their evolution (Boldt 1987; Schwartz 1990; Matt and Fabian 1994; Gendreau 1995). For high redshifts ($1 < z < 5$), this structure would appear in the 1-3 keV band, a regime that is complicated in imaging instruments by M-shell spectral distortions from the gold coating used in grazing incidence X-ray telescopes. Being devoid of such complications, HUBE-CUBIC is an ideal instrument for carefully examining CXRB spectral signatures of iron K-line emission by high redshift extragalactic sources. Such measurements require long exposures averaging over large solid angles to measure deviations of a few percent from the CXRB continuum. This is a challenging observation, but one with rich scientific rewards: direct measurement of AGN evolution from their composite spectrum.

Now let me turn to another topic, black holes, which are regarded in some quarters as a candidate "hiding place" for at least some of the missing baryonic dark matter. However, I will discuss the *physics* of black holes, rather than their potential significance as dark-matter candidate.

BLACK HOLES

Observational astronomers are often nonplussed when General Relativists assert that we can learn no black hole physics from observations of black holes. The General Relativists are correct, however, and the reason is both simple and profound. Black holes are *deeply* understood. Let me demonstrate how simple they are by deriving their existence, assuming that the reader already knows about tensors. Even if the reader does *not* know about tensors (simple enough mathematics!) my point, which is the utter simplicity of black holes, should be proven to the reader's satisfaction. Tensors are sets of functions of the coordinates that are in a sense coordinate-independent and hence are suitable for expressing laws of physics. One simple but very important example of a tensor is 0, which is certainly independent of coordinates! The set of functions $g_{\mu\nu}$ describes a geometry, and is called the metric tensor. The subscripts μ and ν go from one to four, because the Universe has three space dimensions and one time dimension, which totals four. Einstein's fundamental idea was that geometry (our set of functions) is equal to mass-energy, which is zero outside a star: for example, anywhere in the solar system. So, one's first guess at the law of gravitation would be $g_{\mu\nu} = 0$. However, that is clearly no

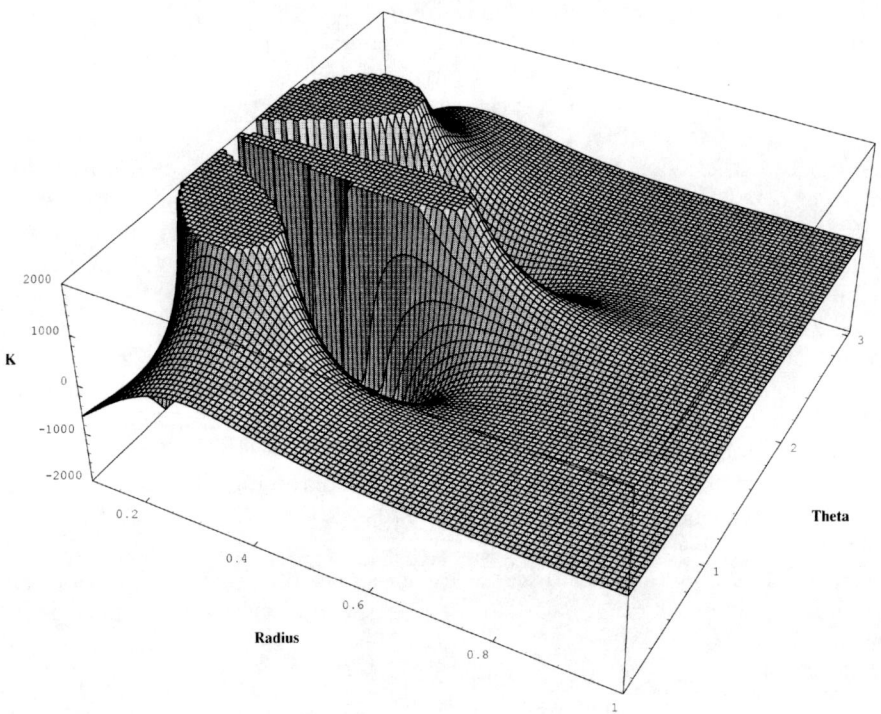

FIGURE 5. A conceptual photograph of a black hole, showing how curved the black hole is as a function of distance from the black hole ("Radius") and North Polar Angle ("Theta"). This particular plot is for a value of $a = 0.6$ (angular momentum per unit mass) and for electric charge $Q = 0.2$. Real black holes in the Universe are unlikely to be electrically charged.

geometry at all. Thus, if one is driven to pursue Einstein's idea, it is necessary to find the next most complicated tensor that involves only the metric tensor. It takes only a page of effort to find $R^\alpha{}_{\beta\gamma\delta}$, and one's next guess clearly would be to set this *new* tensor equal to zero. However, our new tensor is a very famous tensor, the Riemann tensor, and when it is zero, the spacetime involved is defined to be flat. (That, of course, is because it *is* flat for spaces and spacetimes we normally consider to be flat.) So, our second guess is a dismal failure. However, it is possible to form a lower-order (ie, fewer indices) tensor from any tensor by a purely mathematical process called contraction, which involves setting a raised and lowered index equal to each other and summing, to form a new set of (fewer) functions. For example, let us contract the Riemann tensor: $\Sigma R^\alpha{}_{\beta\alpha\delta} = R_{\beta\delta}$. Our third try, of course, is $R_{\alpha\beta} = 0$, ... which are the vacuum field equations of General Relativity. Trivial!

Einstein felt that these equations were too complex to ever allow an analytic solution, but he was very wrong: within six months, the father of my friend the late Martin Schwartzchild found the famous Schwartzschild solution:

$$ds^2 = \frac{1}{1 - \frac{2m}{r}} dr^2 + r^2 d\theta^2 + r^2 sin^2\theta d\phi^2 - (1 - \frac{2m}{r})dt^2 \tag{1}$$

This is one of the most astounding equations in all of physics. The coefficients of the squared coordinate differentials are the four non-zero components of the metric tensor. Schwartzschild's metric differs from the metric of plain ordinary flat spacetime only in the presence of the parameter m, which is the mass of the sun (about a kilometer in appropriate units). This equation predicts Newtonian gravitation as its first approximation, and predicts the precession of the perihelion of Mercury (which was previously unexplained) and also predicts (and *this* prediction you can *see*) the possibility of black holes: you will notice that if distance from the central mass m is $r < 2m$, the negative sign of the final term (time) changes from negative (which is what identified it as *being* time) to positive, while at that same distance the first term, which had been distance from the sun, becomes negative. The two coordinates switch roles! You cannot go back, because you cannot go back in time.

There is nothing more to it than that!

Real black holes in the Universe (black holes that might make up the baryonic dark matter) are, however, not Schwartzschild black holes. They are surely Kerr black holes, which have the added attribute of angular momentum (and, in principle, also electric charge, although real black holes would easily electrically neutralize by strongly attracting the needed charged particles from their environment.)

A picture of a general black hole is shown in Figure 5. What is plotted is the curvature of spacetime, as a function of distance ("Radius") from the rotating and electrically-charged black hole, and of north-polar distance ("Theta"). The quantity plotted is the Kreutschmann scalar K, about which a word is necessary.

If we perform our operation of contraction on $R_{\alpha\beta}$ (after, of course, raising one index) we get $R^\alpha{}_\alpha = R = 0$, where the final step follows from the vacuum field equations. The gaussian curvature R of spacetime around a black hole is zero! That is in contrast to, for example, a beachball, which has gaussian curvature

$2/a^2$, where a is the radius of curvature of the beachball. However, there is *another* scalar which can be formed from (multiple) contraction of the Riemann tensor: $K = R^{\alpha\beta\gamma\delta}R_{\alpha\beta\gamma\delta}$, and K is non-zero for a beachball ($4/a^4$) *and is also non-zero for black holes*. In Figure 1 you are *really seeing* a black hole, in just as intellectually-meaningful a sense as you have ever seen a beachball.

I have never seen a plot of K in any book on General Relativity, and also, I have never seen the algebraic expression for it published, so: I publish it here:

$$\begin{aligned}K = \quad & \frac{m^2/2}{(r^2 + a^2 cos^2\theta)^6}[-30a^6 + 540a^4r^2 - 720a^2r^4 + 96r^6 \\ & + 45a^2 cos2\theta(16a^2r^2 - 16r^4 - a^4) + 18a^4 cos4\theta(10r^2 - a^2) - 3a^6 cos6\theta \\ & - \frac{Q^2 r}{m}(360a^4 - 960a^2r^2 + 192r^4 + a^2 cos2\theta(480a^2 - 960r^2) + 120a^4 cos4\theta) \\ & + \frac{Q^4}{m^2}(42a^4 - 272a^2r^2 + 112r^4 + a^2 cos2\theta(56a^2 - 272r^2) + 14a^4 cos4\theta)]\end{aligned}$$

I do not present the Kerr metric itself, because it is not pretty (for a relatively clean version of the Kerr metric, see Enderlein 1997).

How does it happen that an observational astronomer can be publishing these properties of black holes? The answer is, the miracle of computers. While Mathematica does not "do" tensors normally, I wrote a Fortran program to generate a script that Mathematica could execute. It generates the above equation in about half an hour on my Macintosh G3 Powerbook. We are rapidly entering a new world, where advanced mathematics will be in the hands of "οι πολλοι!"

CONCLUSION

Revolutions are occuring in instrumentation, and in computer manipulation of mathematics and of data. One of the great unsolved mysteries of the Universe is that of dark matter and the intergalactic medium. I have shown how accessible the advanced mathematics of black holes has become, through the advent of powerful handy computers, and I have shown how a quite simple NASA mission could, perhaps, solve the mystery of the dark matter and the intergalactic medium.

REFERENCES

1. Bernstein, R. A. 1998,*PhD Thesis*, California Institute of Technology
2. Boldt, E. 1987,*Physics Reports*, **146**, 215
3. Burrows, D. N. 1996, *CUBIC Instrument Handbook*, The Pennsylvania State University, http://www.astro.psu.edu/xray/cubic/papers/handbook/
4. Enderlein, J. 1998, *Am. J. Phys.*, **65 (9)**, 897
5. Gendreau, K. 1995, *PhD Thesis*, MIT

6. Henry, R. C. 1999, *ApJ (Letters)*, in press
7. Henry, R. C., and Murthy, J. 1994, in *Extragalactic Background Radiation*, ed. D. Calzetti, M. Fall, M. Livio, and P. Madau, Cambridge: Cambridge Univ. Press, 51
8. Matt, G., and Fabian, A. 1994, *MNRAS*, **267**, 187
9. Murthy, J., Hall, D., Earl, M., Henry, R. C., and Holberg, J. B. 1999, *ApJ*, in press
10. Murthy, J., Henry, R. C., Feldman, P. D., and Tennyson, P. D. 1989, *ApJ*, **336**, 954
11. Murthy, J., Henry, R. C., Feldman, P. D., and Tennyson, P. D. 1990, *AA*, **231**, 187
12. Murthy, J., Henry, R. C., and Holberg, J. B. 1994, *ApJ*, **428**, 233
13. Schwartz, D. 1990, *BAAS*, **22**, 1220
14. Sciama, D. W. 1997, *ApJ*, **488**, 234
15. Ueda, Y., Takahashi, T., Inoue, H., Tsurur, T., Sakano, M., Ishisaki, Y., Ogasaka, Y., Makishima, K., Yamada, T., Ohta, K., and Akiyam, M. 1998, *Nature*, **391**, 866
16. Vogeley, M. S. 1998, *BAAS*, **29**, 1207
17. Witt, A. N., and Petersohn, J. K., 1994, in *Proc. 1st Symp. Infrared Cirrus*, eds. R. M. Cutri and W. B. Latter, ASP Conference Series **58**, 91

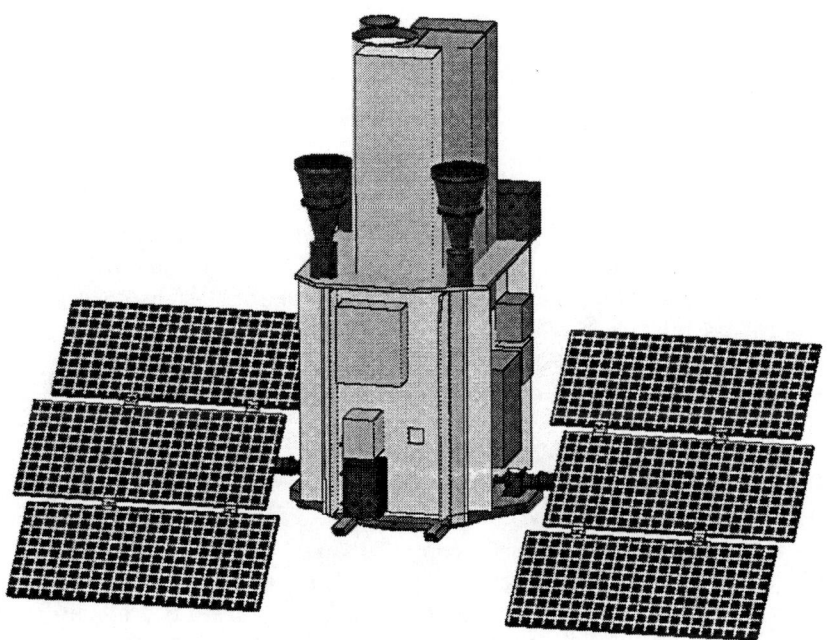

FIGURE 6. The HUBE Instruments mounted in a Ball Aerospace RS2000 spacecraft. The instruments could be used to carry out a thorough all-sky spectroscopic survey of the X-ray and ultraviolet background radiation in about two years. This figure was provided by Mark Skinner.

The Interstellar Medium

Richard C. Henry

Center for Astrophysical Sciences
The Henry A. Rowland Department of Physics and Astronomy
The Johns Hopkins University
Baltimore, Maryland 21218-2686
henry@jhu.edu

Abstract. The study of the interstellar medium has been, and remains, a rich field for exploitation using small missions in the ultraviolet and X-ray spectral regions. I review the history of some such missions (of various sizes), and I also review the capabilities for study of the interstellar medium of *"Hot Universe Background Explorer"* (HUBE), as it was submitted in the second MIDEX round. A new technique for the three-dimensional visualization of the local interstellar medium is also exhibited.

INTRODUCTION

The emission line of Lyman α from cool stars has been available for use in study of the interstellar medium since the rocket measurement of Rottman, Moos, Barry, and Henry (1971). It is remarkable to contemplate the history of that measurement. Paul Patterson had completed a theoretical PhD thesis at Yale, under the direction of Ludwig Oster, examining the possibility of detection of the Lyman α emission line of cool stars by means of rocket observations. The issue was the concern that the stellar Lyman α emission might not be able to traverse interstellar space without being totally absorbed by the interstellar neutral hydrogen, which at that time was believed to have an average density of about 1 atom cm^{-3}. Patterson's rocket, intended to test the technique, was not a success, but Paul's idea stuck in my mind, and when Warren Moos asked me for a suggestion for a star to observe during the engineering gyro correction which was to occur at the beginning of a rocket flight to carry out UV spectroscopy of planets, I suggested Arcturus. We obtained a clear detection of Lyman α emission from Arcturus (following which the mission failed totally), and the chain from that observation, to the Johns Hopkins FUSE mission, is unbroken. Great oaks do indeed from little acorns grow! I have reviewed elsewhere (Henry et al. 1986) the succession of our subsequent observations of Lyman α emission from cool stars with, first, *Copernicus*, and then IUE: both of which were small (or medium-sized) space satellite missions. My interest quickly centered on the problem (which was pointed out to us by Don York) of determination of the cosmologically-important deuterium-to-hydrogen ratio. In Figure 1, I show our *most recent* results that have flowed out of that brilliant initial observation, our high-resolution HST observation of the emission line profile of (as one example) 31 Com (Dring et al. 1997). The deduced chromospheric emission line is shown as a thin line, which is cut in the middle by very strong absorption by the interstellar medium's neutral hydrogen and deuterium. The absorption, while strong, is of course not as strong as had been initially feared.

The interstellar deuterium which we observe must be of primordial origin, perhaps replenished by infall from high-velocity clouds. Linksy et al. (1996) have

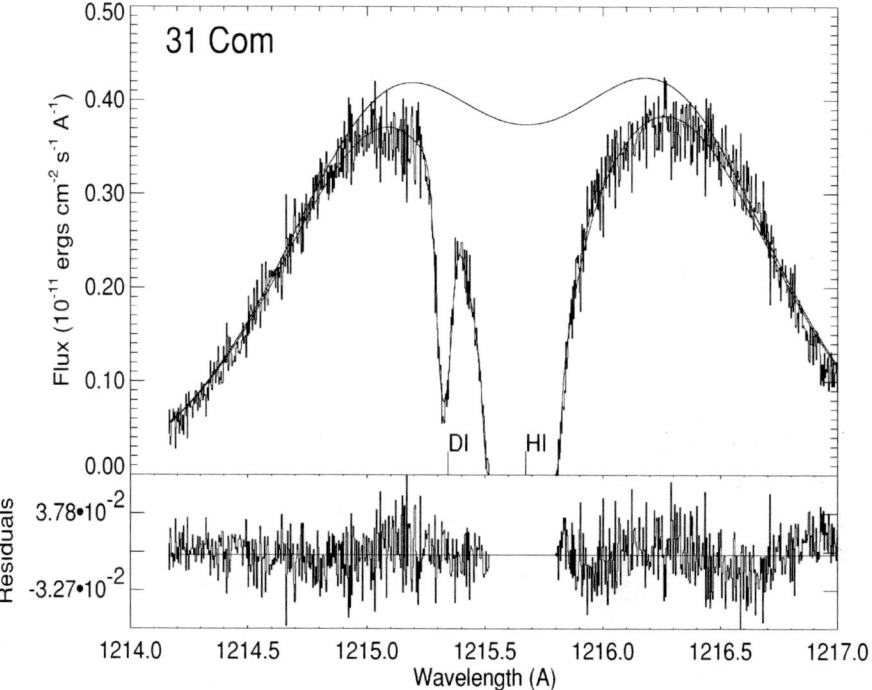

FIGURE 1. Interstellar absorption of hydrogen and deuterium Lyman α, seen against the chromospheric Lyman α emission line of the cool star 31 Com. While the deuterium-to-hydrogen ratio is of great interest for cosmology, it is also of great interest for study of the destruction of deuterium in stars as the galaxy evolves. The column density of hydrogen is, of course, also itself of great interest, for study of the interstellar medium; and deuterium observations of the column density are more secure than are hydrogen observations, because the deuterium line is so much weaker than is that of hydrogen itself. Our measurements of the column density of hydrogen, in the direction of many stars, form the raw material for the construction of three-dimensional maps of the structure of the local interstellar medium, as appears in Figure 3.

especially empasized the importance of these deuterium observations for study of the interstellar medium, since the data analysis is so much more certain for the deuterium than it is for the highly-saturated hydrogen line.

THE INTERSTELLAR MEDIUM

Thus we have seen an almost perfect example of the role of small missions, which fill in from the initial discovery, via sounding rockets, to the fully-mature observations using, in this case, the Hubble Space Telescope. One result of this chain of missions, is new understanding of the interstellar medium. In particular, once a sufficient number of lines of sight have been examined to determine column density, a three-dimensional map of the local interstellar medium can be constructed.

I introduce here a new technique for visualizing the results of such study of the interstellar medium, the use of "crossed-eye" stereoscopic images. In Figure 2, so as to allow the reader to practice the technique with a simple but vivid example, I show "snowflakes" that I have created. All that is necessary is that the reader cross his or her eyes, and superimpose the two images. A very clear three-dimensional image should jump out. Russ (1995) tells us "in many of the images [in this chapter], two adjacent images in a rotation or pseudo-rotation sequence can be viewed as a stereo pair. For some readers looking at them will require an inexpensive viewer which allows one to focus on the separate images while keeping the eyes looking straight ahead (which the brain expects to correspond to objects at a great distance). Other readers may have mastered the trick of fusing such printed stereo views without assistance. Some, unfortunately, will not be able to see them at all. A significant portion of the population seems not to actually use stereo vision, due for instance

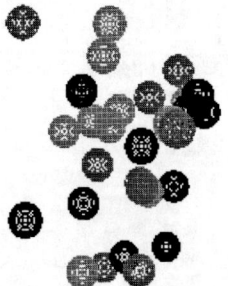 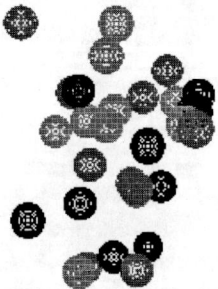

FIGURE 2. Three-dimensional image: cross your eyes, and superimpose the two images. A vivid three dimensional image of snowflakes should materialize. This figure is an "engineering test;" the technique is applied to visualization of the local interstellar medium in Figure 3. The two images need to be reversed if one wishes to use a stereoscopic viewer in place of the crossed-eye technique.

to uncorrected amblyopia ("lazy eye") in childhood." What Russ fails to point out is that the images that are needed for viewing with a viewer are not the same as the images that are needed for viewing with eyes crossed. You may see the difference at http://msx4.pha.jhu.edu/khuuleDir/stereo.html, where you will find a color version of Figure 2, plus the same figure done so as to be viewable correctly with a stereo viewer. The difference in the images becomes quite clear, when you view the latter with the crossed-eye technique.

In Figure 3, I use this new technique to provide a three-dimensional preview of the local interstellar medium. I call it a preview, because the data set on which it is based have not yet been published (Linsky et al. 1996), and so the representation must be regarded as provisional only. Nevertheless, the figure clearly indicates the potential of the technique for three-dimensional visualization. The Sun is located in the Local Interstellar Cloud (LIC), which has two components with slightly different motions. This cloud is partially ionized and is at a temperature of about 7000K. The local cloud is surrounded by a highly ionized bubble, which extends to

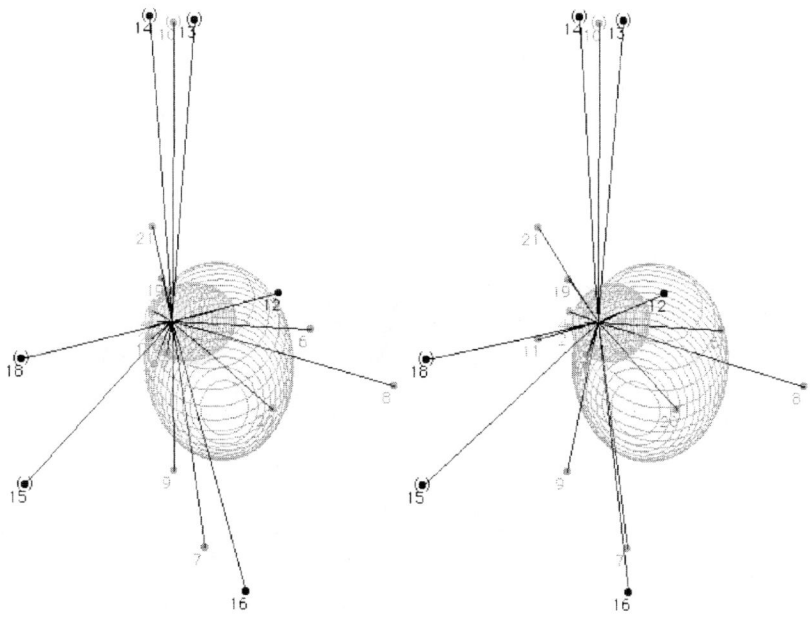

FIGURE 3. Three-dimensional image: a provisional image of the local interstellar medium. Numbered lines-of-sight to various stars are shown (for example, 31 Com is #10). The North Galactic Pole is at the top of the figure, and the Galactic Center lies to the left. Two overlapping local clouds of interstellar gas are seen. The properties of the clouds have been deduced by Linsky, Piskunov, and Wood (1996) from the observed column densities on the various lines of sight that are shown.

more than 100 pc, depending on direction.

FUTURE MISSIONS

Let me highlight, as an example of possible future missions, *"Hot Universe Background Explorer,"* HUBE, as it was submitted to the second NASA MIDEX round. In that submission, thanks to the kind cooperation of Wilt Sanders and his collegues, I was joined by John M. Harlander, Jeffrey J. Bloch, and Barham W. Smith, with whose crucial aid I was able to include HUBE-ISHS (Imaging Spatial Heterodyne Spectrometer) as part of the HUBE payload. The capabilities of HUBE-ISHS for high-resolution spectroscopy of C IV emission from the interstellar medium are shown in Figure 4.

The MIDEX platform is of sufficient capacity that the HUBE proposed mission could also include X-ray spectroscopy of the interstellar medium. To achieve this,

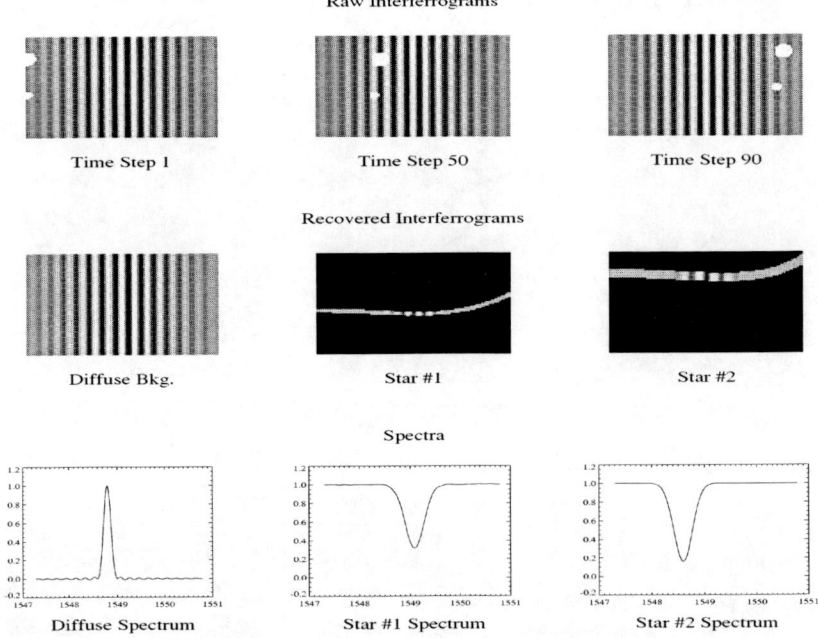

FIGURE 4. HUBE-ISHS carries out high-resolution spectroscopy of the diffuse C IV radiation in the galaxy. With no de-convolution required, there is clean separation of the emission spectra of point sources and the diffuse background. This figure was created by John M. Harlander. The instrumental technique is described by Harlander (1991), Harlander and Roessler (1990), and Harlander, Roessler, Reynolds, Jaehnig, and Sanders (1993). The spectral resolution of HUBE-ISHS is 15 $km\ s^{-1}$ which is better than IUE high resolution, and comparable to HST-GHRS.

I was joined by David N. Burrows, the Principal Investigator for CUBIC (Burrows 1996). Figure 5 gives the reader some idea of the remarkable capability of CUBIC for providing high-resolution spectroscopy of the interstellar medium. I have presented the other ultraviolet spectroscopic capabilities (Table 1) of HUBE elsewhere (Henry 1999a). Also, the synergism of the HUBE ultraviolet spectroscopy with ISHS and CUBIC for analysis of the Cygnus Loop nebula, as an example, has been comprehensively illustrated in Henry (1999b).

CUBIC is an ideal choice for the first all-sky high-spectral-resolution X-ray spectroscopic survey, because of its high spectral resolution, its simplicity, and its ro-

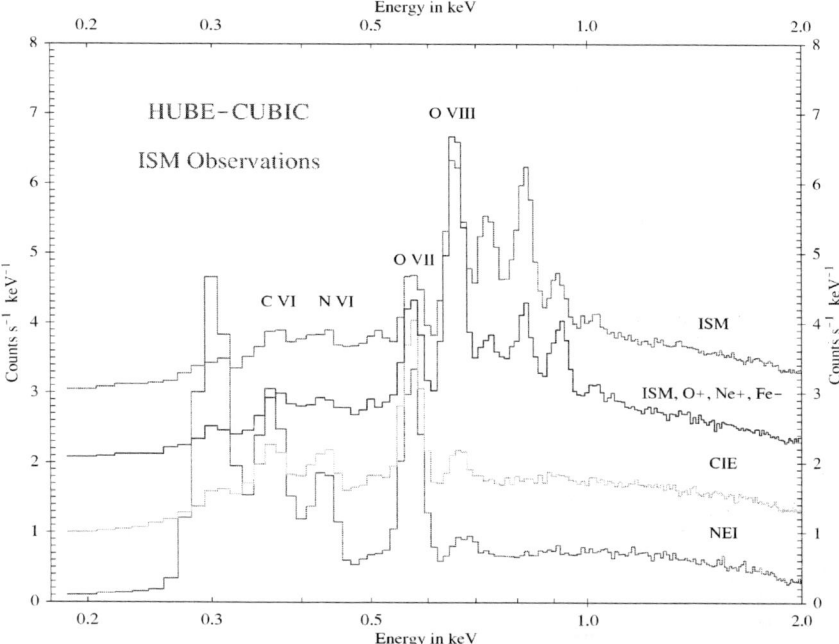

FIGURE 5. HUBE-CUBIC carries out high-resolution X-ray spectroscopy of the diffuse radiation of the galaxy. I have created this figure from simulations that were provided by David N. Burrows. The top curve shows CUBIC response to a standard two-component model of the ISM, based on proportional counter fits to the high latitude diffuse X-Ray background (Nousek 1978). The second curve is identical except that the abundances of O and Ne have been increased by 50% while that of Fe has been reduced by half, resulting in large changes in the relevant lines. The third spectrum is for a single temperature, solar abundance, collisional ionization equilibrium (CIE) model fitted to CCD sounding rocket data for a similar high latitude direction, while the bottom spectrum is for the same model, but in nonequilibrium ionization (NEI) with an ionization parameter of $\log(nt) = 10.0$. CUBIC will be able to measure abundances to 25% accuracy and will readily distinguish betwee CIE and NEI conditions, as well as between single and multiple-temperature plasmas.

TABLE 1. Instruments that were proposed for the second MIDEX submission of HUBE.

Instrument	$\lambda\lambda$ (Å)	Prime Science Goal
EUVS (Extreme-Ultraviolet Spectrometer)	850 - 1200	Map O VI and UV bkgd
LαS (Lyman α Straddle Spectrometer)	850 - 1400	Detect UV bkgd step at 1216 Å
ISHS (Imaging Spatial Heterodyne Spectrometer)	1544 - 1568	*Velocity-resolved* C IV maps
CUBIC (X-ray Spectrometer)	1 - 250	X-ray background spectrum
FUVS (Far-Ultraviolet Spectrometer)	1230 - 1800	High-latitude UV bkgd
UVI (Ultraviolet Imager)	1350 - 2000	Map the diffuse background

bustness. CUBIC is significantly more sensitive than any previous X-ray mission except HEAO-1, and CUBIC provides an 8-fold increase in spectral resolution over HEAO-1.

While it *is* possible to achieve *still higher* spectral resolution by use of a

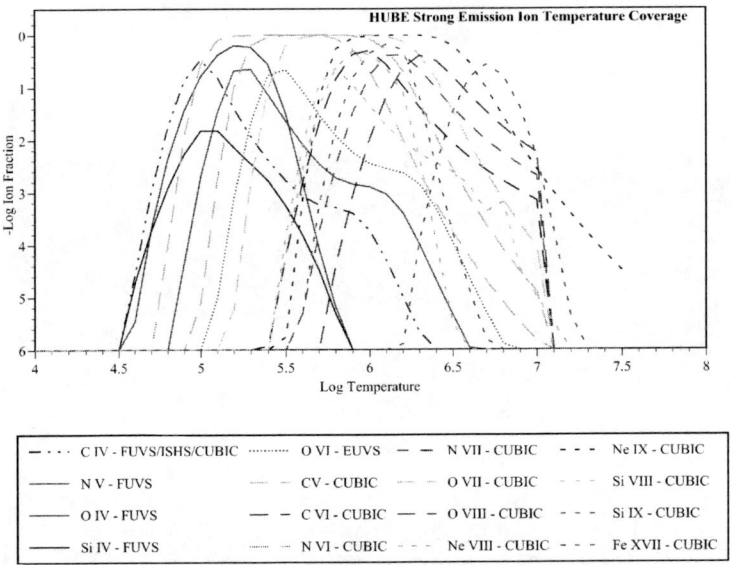

FIGURE 6. HUBE, as proposed in the second MIDEX round, carried a range of instruments that have the spectroscopic capabilities that are summarized in this figure by Jeffrey J. Bloch. HUBE-ISHS and HUBE-CUBIC have been described to some extent in the present paper. HUBE-EUVS and HUBE-FUVS are described by Henry (1999a). The capabilities of all of these instruments together to provide comprehensive analysis of the Cygnus Loop nebula (as an example) is described by Henry (1999b). As you can see from the present figure, little of the interstellar medium could escape HUBE's spectroscopic scrutiny.

cryogenically-cooled detector, such a mission has limited lifetime and would be much more effective as a follow-up mission to CUBIC, to resolve any regions that remain spectroscopically-confused. Such regions should be few; see Figure 5.

The exceptional completeness of HUBE as a mission for spectroscopic analysis of the interstellar medium is demonstrated in Figure 6, which shows the ion temperature coverage of the complement of instruments (Table 1) that made up HUBE in its second MIDEX incarnation. Particularily important is the fact that several ionization stages of important elements are included, and that ions (such as O VI) are observed in emission, that are also observed in absorption by other planned missions (in the case of O VI, the FUSE mission).

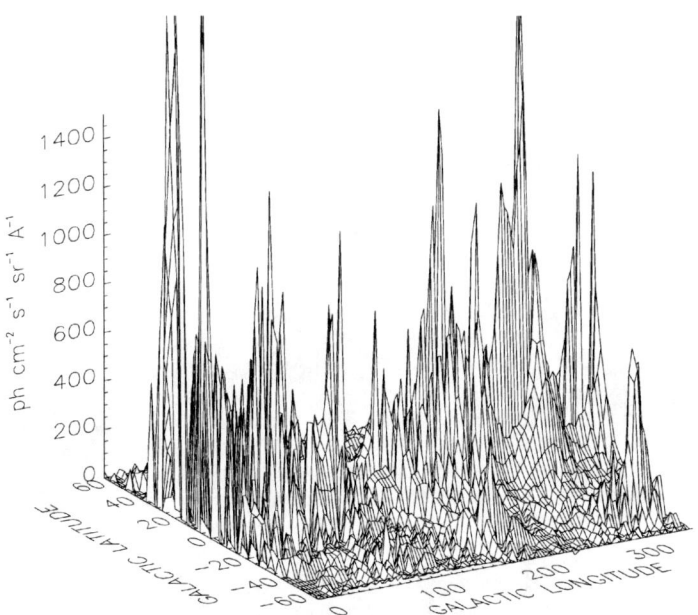

FIGURE 7. Model (Murthy and Henry 1995) of the expected distribution of ultraviolet light that has been scattered once from interstellar dust. The dust is assumed to be very strongly forward-scattering in this model, which explains the patchy appearance: individual stars can create patches of forward-scattered light. Attempts to determine the albedo and scattering parameter of the dust grains are clearly doomed to failure, if they rely on a few scans as a function of galactic latitude. A detailed map of the sky in the ultraviolet, however, could unambiguously determine the albedo and scattering parameter.

INTERSTELLAR DUST

My final topic is interstellar dust, and the profound role it plays, not only in astrophysical processes, but in hindering our observation of the interstellar medium. Two views of strongly-forward-scattering interstellar dust as viewed from our location in the galaxy are given in Figures 7 and 8. The two figures are different representations of the same data, namely one of the models of dust-scattered starlight of Murthy and Henry (1995). The particular model that I have chosen for display is one I believe to be close to the truth: one in which the Henyey-Greenstein scattering parameter g = 0.9, corresponding to very strong forward scattering.

Note that if this model provides an accurate representation of what really occurs, the pattern of scattered starlight over the sky is very asymmetric. This partly reflects the asymmetric pattern of the ultimate, underlying, source of the radiation, the UV-emitting stars, which are located mainly in Gould's belt; but also reflects the very strong absorption of light from more distant stars (which are distributed

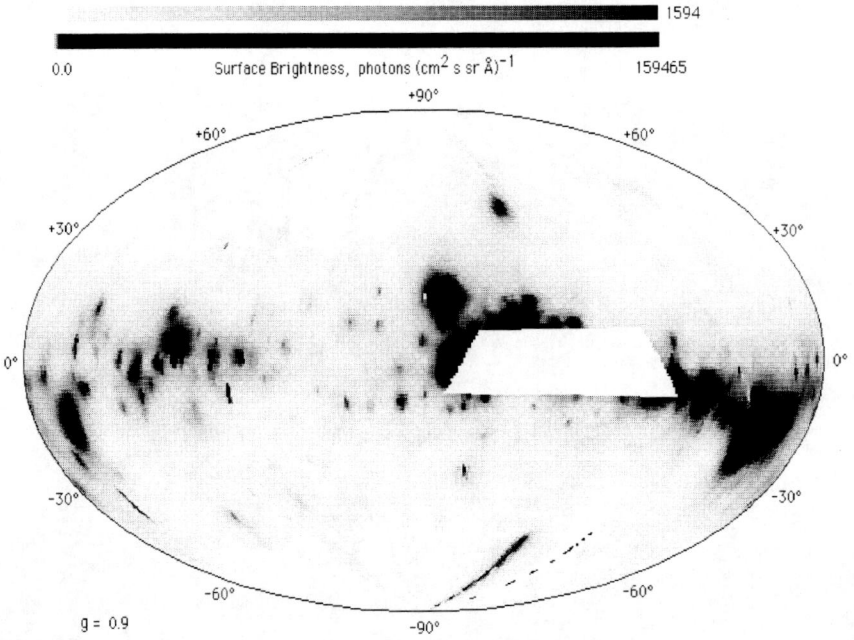

FIGURE 8. Model of the expected distribution of ultraviolet light that has been singly-scattered from interstellar dust. This is a different presentation of the same model that is shown in Figure 7. In the present representation, the North Galactic Pole is at the top of the figure, and the Galactic Center is at the center of the figure. Galactic longitude increases to the left, as you look at the figure. Gould's belt is tipped 19° with respect to the Galactic Plane. A region where no H I data were available is left blank. The model is graced with a few artifacts.

much more symmetrically with respect to galactic coordinates.) The region in the Galactic Plane that is dimmest in Figure 8 is also nearly devoid of direct UV light from stars, as may be seen in Murthy and Henry (1995): they show a model, plus an integration of the starlight that was actually observed in the TD-1 mission.

The correct values of for albedo and for the Henyey-Greenstein scattering parameter in the ultraviolet are not well established yet. The situation has been reviewed by Bowyer (1991); see also Witt et al. (1992), Witt and Petersohn (1994), and Henry and Murthy (1993). It is clear from Figures 7 and 8 that what is needed to resolve all issues, is an all-sky map of the distribution of the scattered starlight in the ultraviolet. It should be entirely possible to unambiguously assign values to the albedo and scattering parameters from such a map, assuming, of course, that these parameters have a unique value, which is not likely to be the case. Regardless of the latter concern, an all-sky map of the UV scattered light is clearly a high priority.

REFERENCES

1. Bowyer, S. 1991, *ARAA*, **29**, 56
2. Burrows, D. N. 1996, *CUBIC Instrument Handbook*, The Pennsylvania State University, http://www.astro.psu.edu/xray/cubic/papers/handbook/
3. Dring, A. R., Linsky, J., Murthy, J., Henry, R. C., Moos, W., Vidal-Madjar, A., Audouze, J., and Landsman, W. 1997, *ApJ*, **488**, 760
4. Harlander, J. M. 1991, *PhD Thesis*, University of Wisconsin
5. Harlander, J. M., and Roessler, F. L. 1990, *Proc. SPIE*, **1235**, 622
6. Harlander, J. M., and Roessler, F. L., Reynolds, R. J., Jaehnig, K, and Sanders, W. 1993, 1990, *Proc. SPIE*, **2006**, 310
7. Henry, R. C., Murthy, J., Moos, H. W., Landsman, W. B., Linsky, J., Vidal-Madjar, A., and Gry, C. 1986, *ESA Proceedings of an International Symposium on New Insights in Astrophysics*, London: University College London, 555
8. Henry, R. C. 1999a, *Multifrequency Behaviour of High Energy Cosmic Sources*, Memorie della Societa Astronomica Italian, eds. F. Giovannelli and L. Sabau-Graziati, in press
9. Henry, R. C. 1999b, *IAU Colloquium No. 171, The Low Surface Brightness Universe*, ed. J.I. Davies, C.D. Impey, and S. Phillipps, ASP Conference Series, in press
10. Henry, R. C., and Murthy, J. 1993, *ApJ (Letters)*, **418**, L17
11. Linsky, J. L., Piskunov, N., and Wood, B. E. 1996, *unpublished*
12. Murthy, J., and Henry, R. C. 1995, *ApJ*, **448**, 848
13. Nousek, J. 1978, *PhD Thesis*, University of Wisconsin-Madison
14. Rottman, G. J., Moos, H. W., Barry, J. R., and Henry, R. C. 1971, *ApJ*, **165**, 661
15. Russ, J. C. 1995, *The Image Processing Handbook, Second Edition*, Boca Raton: CRC Press, 621
16. Witt, A. N., Petersohn, J. K., Bohlin, R. C., O'Connell, R. W., Roberts, M. S., Smith, A. M., and Stecher, T. P. 1992, *ApJ*, **395**, L5
17. Witt, A. N., and Petersohn, J. K., 1994, in *Proc. 1st Symp. Infrared Cirrus*, eds. R. M. Cutri and W. B. Latter, ASP Conference Series **58**, 91

SIXE: An X-ray experiment for a minisatellite

Jordi Isern*†, Eduardo Bravo*‡, Jordi Gómez-Gomar*, Margarida Hernanz*†, Enrique García-Berro*‡, Franco Giovannelli‖, Cesare D. La Padula‖, Lola Sabau¶, Jordi Gutiérrez*‡, Jordi José*‡, Domingo García-Senz*‡, Joan Bausells§, Joan Cabestany*‡, Jordi Madrenas*‡, Manuel Angulo¶, Manuel Fernández-Valbuena¶, Erardo Herrera¶, Manuel Reina¶, Antonio Talavera¶

*Institut d'Estudis Espacials de Catalunya (IEEC) [1]
Gran Capità 2-4, 08034 Barcelona, Spain
†Consejo Superior de Investigaciones Científicas (CSIC), Spain
‡Universitat Politècnica de Catalunya (UPC), Barcelona, Spain
‖Istituto di Astrofisica Spaziale (IAS-CNR), Frascati, Italy
¶Instituto Nacional de Técnica Aeroespacial (INTA), Torrejón de Ardoz, Spain
§Centro Nacional de Microelectrónica (CNM, IMB-CSIC), Bellaterra, Spain

Abstract. *SIXE* (Spanish Italian X-ray Experiment) is an X-ray detector with geometric area of 2300 cm^2, formed by four identical gas-filled Multicell Proportional Counters, and devoted to study the long term spectroscopy of selected X-ray sources in the energy range 3-50 keV. The main characteristics of *SIXE* are: time accuracy of 1 microsecond, spectral resolution of 5% for $E > 35$ keV and $46/\sqrt{E}\%$ for $E < 35$ keV, continuum sensitivity (3σ in 10^5 s) of 2×10^{-6} ph/cm^2·s^{-1}·keV^{-1}, and line sensitivity (3σ in 10^5 s) of 8×10^{-6} ph/cm^2·s^{-1}. The size of the instruments and the requirements of the payload (weight 103 kg, full dimensions $660 \times 660 \times 450$ mm^3, power budget < 60 W, on-board memory 2 Gbits, telemetry rate < 100 kbps) make this experiment fully compatible with a minisatellite mission. The experiment, whose feasibility study has just been finished, has been proposed for flying on the Spanish MINISAT-02 satellite, in a 3 years long mission starting about 2002-2004. The main scientific goal is the study of the short and long term variability of a selected set of X-ray sources, such as quasars, Seyfert galaxies, high and low mass X-ray binaries, etc. The philosophy of the mission will provide the unique opportunity for the study of X-ray sources with a temporal accuracy of 1 microsecond all through the time range $10^{-5} : 10^7$ s.

---

[1] http://www.ieec.fcr.es/

INTRODUCTION

During the last years, X-ray Astronomy has experienced a large development and has significantly contributed to a better understanding of our Universe. Nevertheless, in spite of the success of X-ray Astronomy, there still exist today large areas where we lack a profound knowledge of the detailed physical mechanisms responsible for the emission of X-rays. These areas of course require new methods of research and, consequently, new instruments. One of these areas is, for sure, the study of the temporal variability of X-ray sources, especially those which radiate in the region of high energy, namely the $3 \leq E$ (keV)≤ 50 band. This band is especially important because some of the most interesting X-ray sources emit a sizeable fraction of their energy in this spectral region, besides being variable sources. The underlying physical mechanism powering most of these sources is, roughly speaking, accretion onto a compact object (white dwarfs, neutron stars or black holes). Moreover, most of the information we can retrieve from these systems comes from a detailed analysis of their variability (QPOs, pulsations, bursts,...) and of their spectral features. To this point, it is important to remark that the variability on time scales of the order of a millisecond and even less can provide us with valuable information about the behavior of matter in the surface of a neutron star or close to the horizon of a black hole.

One of the major drawbacks of large missions is the impossibility of devoting a large fraction of the observing time to the continuous monitoring of a single source, since most of these facilities have a significant observing pressure and, besides, sometimes there are conflicts between the different experiments on board. Nonetheless there are several astronomical problems that can be only solved through a continuous monitoring of the relevant sources or, at least, through long observing runs. Such large observing times can only be achieved through a small mission, as it is the case of a minisatellite, in which apparently major drawbacks, such as its small size and cost, can be transformed into advantages. As an example, *RXTE* on average provides only 50 ks of effective observing time to any source. The small mission presented here (Spanish-Italian X-ray Experiment, *SIXE*) could provide almost 200 times more coverage. Obviously, this can only be done if a small number of sources is selected in a very careful way. To be precise, *SIXE* could provide the unique possibility of observing phenomena with time scales within the range 10^{-5} to 10^7 seconds, something totally unprecedented (the experiment USA on board the *ARGOS* satellite gives similar performances, but observations will be affected by large gaps due to the polar orbit of the satellite, see the paper by K. Wood in this same volume).

In the next paragraphs we mention just a few examples of which kind of science can be done with *SIXE*. The reader will find a thorough description of such science goals in the next section. The most important topics on which *SIXE* could cast some light are the following:

1. Analysis of the time variability on time scales of a few tenths of millisecond of galactic sources in several evolutionary stages. This could provide valuable information about the behavior of matter in the neighborhood of the surface of a neutron star or the neighborhood of the event horizon of a black hole.

2. Very recently it has been discovered that some of the X-ray pulsars which have an accretion disk around them present phases during which the spin velocity increases followed by a sudden change to a phase in which the pulsar brakes down. Until now it has been completely impossible to resolve the transition with high temporal accuracy. *SIXE* could do that in a few cases.

3. The current theories predict that in some clusters of galaxies we should observe an excess of non-thermal emission, which could provide information about the inter galactic magnetic field. Provided its large integration times, *SIXE* could measure this effect.

The total budget for *SIXE* is estimated to be 4,019 kECUS, or about 671 MPts. In this estimate we include the costs of design, development, checking and quality control of the instrument, its integration and qualification, the implementation of the ground and user segment, the operational phase and the management costs from the design phase to the operational phase. Under no circumstances major changes in the MINISAT-01 platform are required and only small modifications are planned. *SIXE* has been designed to be fully compliant with the basic characteristics of the MINISAT program and it is only necessary to minutely exhaust the already attainable possibilities, mainly in attitude control, since we require a precision of 0.15 deg.

SCIENTIFIC GOALS

Previous space-borne X-ray instruments have shown that many sources emit appreciably in the hard X-rays domain and that most of them show variability. These sources are mainly related to accreting compact objects, both galactic and extragalactic. Most valuable information about these phenomena is gained through the study of their variability signatures (bursts, pulsations, QPOs, etc.) and of their spectral properties. All major hard X-rays experiments share a limitation of observation time, since they have to satisfy a great number of observers and often the requirements of many instruments. Therefore, long duration observations of individual objects are excluded of their observing programmes. However, plenty of questions about hard X-ray sources require either long-term monitoring or very long integration times. Such type of observations will be possible for a small mission devoted to a reduced number of targets ($\sim$ 20 targets during accumulated periods of $\sim$ 2 months), like *SIXE*, which in addition offers an excellent time resolution and accuracy. *SIXE* is designed to study variabilities in the unprecedented range of time scales going from 10^{-5} to 10^7 s, covering from the characteristic timescales of instabilities related to accretion onto neutron stars up to those of AGN variability. Simultaneous coverage in the optical will be possible thanks to the inclusion of an Optical Monitor Camera (OMC). Potential targets for *SIXE* are further commented in the following subsections.

Galactic X-ray Sources

Cataclysmic Variables

Long duration observations of hard X-ray emission by *SIXE* with simultaneous coverage in the optical would be of much interest to address some of the open issues concerning X-ray emission from non-magnetic CVs: morphology of the outbursts in X-rays as compared with the optical, comparative study of X-ray emission during quiescence and outburst (in order to better understand the mechanisms of emission), correlation between rapid oscillations in the optical and in X-rays (in order to constrain the location and nature of the pulsation). On the other hand, the temporal evolution of the X-ray emission of many magnetic CVs, observed with *ROSAT*, *GINGA*, *ASCA* and *RXTE*, reveals complex behaviors changing considerably with time. Thus again, long time observations of some selected magnetic CVs with good temporal resolution, would be of great interest to shed light to the complex behavior of these systems.

Pulsating High and low-mass X-ray binaries

Pulsars associated to Be stars. The X-ray activity in these systems is generated as a consequence of the interaction of the compact object with a radial outflow. The most characteristic features of the X-ray light curve are the occurrence of a series of outbursts and the presence of occasional giant ones. The simultaneous measuring of the X-ray and the optical emission during all the cycle would be of much interest, in order to understand the interaction between the neutron star and the Be star. Also, a systematic monitoring of these sources could help to elucidate if a transient disk forms. This type of observations will be possible for *SIXE*.

Pulsars associated with early supergiants. A detailed study of the Fe-line with respect to the continuum X-ray radiation at different orbital phases should provide precious information on the line formation region. Only *SIXE* will be able to measure the emission of Vela X-1 like systems during several orbital phases since the orbital period is ~ 9 days.

The long-term spin evolution of pulsars. The continuous pulse monitoring from *BATSE* of several X-ray pulsars has shown that they present intervals of spin-up followed by intervals of spin-down, with nearly equal torques, which is very strange. Two important questions are why the magnitude of the spin-up and spin-down torques are so similar, and what determines the reversal time scale of the torques. From the observational point of view two things must be taken into account:

1. It has not been possible to determine the torque on short time scales. Transitions between spin-up and spin-down in Cen X-3 occur on timescales smaller than 10 days, that *BATSE* was not able to resolve. On the other hand, Her X-1 was sampled infrequently at 35 days intervals.

2. The bimodal torque behavior demands step function-like changes in the mass accretion rate which, in turn, must be due to changes in the star or in the accretion disk.

Therefore, it should be of the highest interest to resolve in X-rays and in the visible the transition between the spin-up and the spin-down phases. Since the counterparts of Cen X-3 and Her X-1 have magnitudes in the range V=13-15 and the OMC takes 100 s to reach V=18, it is clear that it would be possible to obtain a good optical coverage of these sources.

Non-pulsating X-ray binaries

SIXE can provide information on the following issues:

- The orbital period of many LMXBs is not well known: the long term monitoring of these objects can improve the knowledge of the orbital parameters and validate the existing theoretical models that predict the stability conditions for the disk.

- The long term evolution (weeks, months..) in the colour-colour diagram of Z and atoll sources and different kinds of temporal variability (QPOs, red noise, flares...) correlated with the different spectral states. The eventual correlation between optical and X-ray emission could clarify the problem of the radiating process and further partial reprocessing of the X-ray component into the optical.

- The study of the short timescales, ≤ 0.1 ms, allows to understand the behavior of matter in the vicinity of the neutron star surface: Nuclear burning, non-radial oscillations, rotation, millisecond variability and so on. *RXTE* has discovered kilohertz QPOs in 18 LMXRBs, with two dominant frequencies which maintain their difference in frequency, but in some cases this separation has been observed to change with time. Also, the detection of QPOs along the different states of these sources could provide important information about the question of the peak separation and its relationship with the spin of the neutron star.

Binaries containing a black hole

Simultaneous long term monitoring of both the hard and the ultrasoft components of the X-ray emission, and the determination of the QPO and other temporal variations during the outbursts and quiescent stages, should be of the highest value to understand the behavior of relativistic accretion disks of black hole candidates. A very interesting observation could be the simultaneous measurement of the optical and X-ray light curves in order to obtain the delay in the optical reprocessing. These systems have typical periods in the range of 5-155 hours and the expected delay is of the order of 6-60 s, depending on the inclination of the orbit and the orbital phase. Thus, the temporal resolution of the optical camera should be as short as 1 s. This could allow to obtain the inclination of the orbital semiaxis with a high precision. Other specific objectives of *SIXE* could be the following:

- Frame-dragging in spinning black holes. As in the case of neutron stars, certain types of QPOs in the light curves of several black holes can be due to the relativistic dragging of inertial frames around spinning objects. These QPOs are expected to be very stable and to extend from few Hz to several hundred Hz.

- Soft X-ray transients. Long term monitoring of such objects could provide the most valuable information about the process that triggers the burst.

- Galactic microquasars. *RXTE* observations have consisted in a long term observation with the All Sky Monitor and a series of observations with the PCA during several years, lasting 1 to 10 ksec each one, with periodicities of 1 week, for a total of $\sim$ 1.5 Msec. *SIXE* could provide a more complete coverage in time, a good efficiency above 10 keV, and a better determination of the background. Furthermore, the optical monitor will allow to obtain the optical light curve in two or three bands.

Other sources

Young Open Clusters. SIXE could examine the hard X-ray emission from few Berk87-like clusters contained in the *COS B* error boxes from which gamma-ray emission has been measured with *EGRET* in order to understand the mechanism of such emission.

Pre-main sequence flaring stars. SIXE could monitor a selected group of stars among those producing flares in order to improve the statistics, their role in the global energetic balance and to measure the rotational period of such stars.

Extragalactic X-Ray Astronomy

In the extragalactic sky, Active Galactic Nuclei (AGN) are the prototypical variable X-ray sources and the primary targets for *SIXE* should be Seyfert galaxies, BL

Lacs, and so on. An instrument like *SIXE*'s X-ray detector plus an optical camera able to perform simultaneous observations of AGNs in two bands (or even just at one) for periods of time of the order of weeks would be of the highest interest.

Seyfert 1 galaxies and QSOs

SIXE can provide the definitive confirmation or rejection of the standard picture for the X-ray spectra of Seyfert 1 galaxies, by comparing the variability of the reflection bump, the Fe line, Fe K edge when present and the underlying continuum. Generally speaking, the reflected components should be synchronous and follow variations of the underlying continuum. Encouraging results have been found with *RXTE* observations of MCG-6-30-15.

Seyfert 2 galaxies

The nominal energy resolution of *SIXE* at 6 keV is expected to be around 0.9 keV, which means that the Compton reflected component will be just deblended from the higher energy emission lines, allowing tests of the geometry of the various reflectors involved. The reflecting material is expected to be far from the central source, and therefore not expected to vary on short timescales. Recent *ASCA* observations of the Seyfert 2 galaxy NGC 7172 suggest that the Fe K line and the underlying continuum vary with the same factor on long timescales and also show variations of up to 30% on scales of hours. Observations spread over timescales of months to years could provide much information on the location of the reflecting material, together with possible long-term variations of the absorbing column and the Fe K absorption edge. Furthermore, it seems that the spectrum of some Seyfert 2 tend to flatten at high energies. Long exposures and good sensitivities at high energies could clarify this question.

Narrow line Seyfert 1 galaxies

The role of *SIXE* on these objects would be crucial. As in the case of *RXTE*, it will determine their spectral shape at hard photon energies (hardening at high photon energies has already been reported with *ASCA* observations of NLS1s). But most importantly, it can monitor the variations of the soft and hard components. Variations of the hard component have been reported for NGC4051, Mrk 766 and IRAS 13224-3809. *SIXE* observations will certainly yield information about the fundamental nature of this class of sources.

Clusters of Galaxies

Many observations of clusters of galaxies have been performed in the range of 1-10 keV, but the measurements in harder X-ray regions, where non thermal excesses should show up, are very scarce and only upper limits are presently available. The present limit for this emission in the Coma Cluster was obtained by the $CGRO$/OSSE and is $< 10^{-6}$ cm^{-2} s^{-1} keV^{-1} at 50 keV in the continuum.

In order to detect the hard X-ray excess of these regions, it will be necessary to reach sensitivities better than 10^{-6} cm^{-2} s^{-1} keV^{-1} at energies larger than 40 keV, which can only be obtained with a large area instrument and long exposure observations. On the other hand, a limited FOV, of the order of 1 deg $\times$ 1 deg, will be required in order to avoid source confusion. $SIXE$ offers a unique possibility to fulfill these requirements while, additionally, it provides a good energy response down to 3 keV, necessary to measure the tail of the thermal X-ray emission.

Very recently, Beppo-SAX has observed the Coma Cluster in the energy range of 0.1-10 keV with the MECS instrument and in the range of 15-200 keV with the PDS and detected a clear evidence of an excess of thermal emission above $\sim$ 25 keV, providing the first evidence of the long sought inverse Compton emission on the cosmic background.

MISSION CHARACTERISTICS

The mission has been conceived for a minimum duration of 3 years. The selected orbit should be circular and with a height between 550 and 600 km. The preferred inclination is between 20 deg and 25 deg. These values have been derived from a compromise between the need of minimizing the background radiation, the situation of the Maspalomas (Canary Islands, Spain) ground station, which has been selected as the primary ground station, and the coverage of the appropriate sky areas during adequate observing times, according with the spirit of the mission. The advantage of these orbits is that they are accessible to small launchers and that are very similar to the orbit selected for the MINISAT-01 spacecraft, thus leading to an optimal use of the actual ground station structure. Given the dimensions and the weight of the instruments, a launcher like Pegasus-XL can be used in this mission.

$SIXE$ will observe very bright sources and, therefore, a large amount of data will be generated. The number of daily contacts between Maspalomas and the spacecraft will be of the order of 6 with a total amount of contact time of roughly 50 minutes. Therefore, it seems adequate to have an additional ground station. We have requested the use of the Malindi (Kenia) italian ground station for such a purpose. Should this secondary ground station be available, the total amount of contact time would be of more than 100 daily minutes and the number of useful contacts would be doubled. Alternatively, the ground stations in Perth (Australia) and Kourou can also be considered. With the use of the ground stations in Maspalomas and Malindi, and using the same band as in MINISAT-01, a telemetry

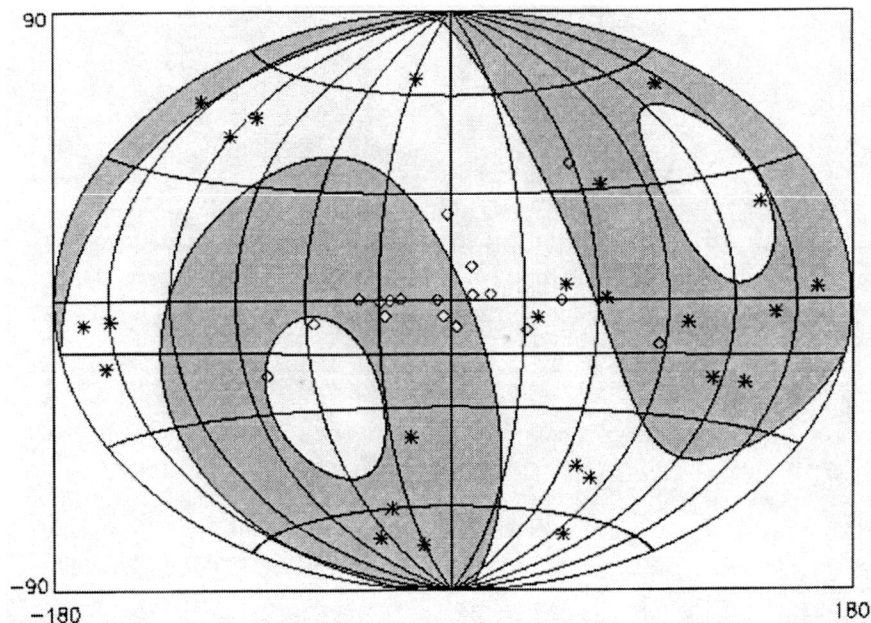

FIGURE 1. Uneclipsed observation bands in galactic coordinates, and a sample of possible targets: galactic (diamonds) and extragalactic (stars).

rate of ~ 60 kbps would be available for *SIXE* (daily average). This telemetry rate is compatible with an event mode data analysis (event time with 1 μs resolution, 256 channels for the event energy) for all but a few of the brightest X-ray sources.

The philosophy of *SIXE* is to observe continuously a few selected sources of each target type. The selected orbit allows for the uneclipsed observation of several sources for periods of up to 18 days (for 30% of potentially interesting sources). In Fig. 1 we have plotted the positions of a sample of possible targets together with the uneclipsed observation bands. The position of the bands will change with time, allowing new targets to be observed continuously during long times. The pointing of the instrument should avoid directions closer than 30 deg to the Sun direction. This requirement still allows the continuous observation of many sources for up to 80 days (again 30% of potentially interesting sources). The fraction of time that will be unuseful for scientific operations due to the crossing of the Van Allen belts (SAA) is lower than 10%.

INSTRUMENTS

The detector to be used in a mission like *SIXE* is naturally chosen from the above mentioned scientific issues, namely the conditions of stability and sensibility to high energy photons. The only detectors that can simultaneously accomplish both requirements are the class of multiwire gas proportional counters. Regarding the need to achieve a high time resolution during long runs (of the order of a tenth of a millisecond during a few years) the best solution is to use an on board GPS receiver. It is important to remark that *SIXE* has a large effective area in the high energy region, comparable to that of more complex missions, which converts *SIXE* in a highly competitive instrument when studying phenomena which show variability. The proposed configuration has four X-ray detectors with an effective area of about 500 cm^2 each one. The optical axis is parallel to the plane of the solar panels. Therefore, we discard a solution with fully orientable solar panels. This condition stems from the inherent costs which are unavoidable with a reasonable budget, and implies that *SIXE* will not be able, generally speaking, to deal with targets of opportunity. However, *SIXE* will have a small freedom to temporarily suspend the planned program in order to observe serendipitous phenomena of high scientific interest. In order to deal with the variability of faint sources or to detect small fluctuations in the X-ray flux, it is of the maximum importance to determine in real time and with the highest accuracy the internal noise of the detectors. This point is so important to us that we have decided to devote two detectors to measure the internal noise. For this reason, the detectors have been placed into two banks with an offset angle of 3.5 deg. This procedure allows the instrument to point simultaneously to the source and to the background. In order to maintain a sustained equivalence between both pairs of detectors, the banks will be regularly switched. Nevertheless the ultimate configuration could be changed during phase B if new arguments relevant to our decision appear. Keeping in mind the ultimate challenge of improving the scientific output of *SIXE* without increasing too much its cost, we have considered the possibility of substituting the optical sensor of the instrument (but not that of the platform) by an optical camera, which could work simultaneously as a sensor and as an optical monitor with two or more photometric filters. Such a solution, which is technically feasible, will make *SIXE* a leading and unique multifrequency instrument.

There will be a total of three instruments working concurrently in the *SIXE* experiment: an X-ray detector, a South Atlantic Anomaly detector (SAAD), and the Optical Monitor Camera (OMC). The X-ray detector will be formed by four identical Multi-Wire Proportional Counters (MWPC). Each module will be filled with a mixture of Xe, Ar and isobutane, and will comprise 70 detection cells (5 rows of 14 cells each) defined by a distribution of cathodes and anodes at a relative potential of $\sim$ 2000 V. There will be 60 cells devoted to the detection of the X-ray photons, 9 cells devoted to veto the particles entering through the detector walls, and one calibration cell. X-ray photons penetrating through the detector window will interact with the gas inside, generating a charge avalanche which will be collected on one of the anodes allowing to determine the energy and time of arrival of the photon. The characteristics of the detector window and the gas mixture have been selected to optimize the efficiency in the nominal range $3 - 50$ keV. The detector efficiency can be seen in Fig. 2 as a function of photon energy. The combination of the four modules will give a total sensitive area of ~ 2000 cm^2,

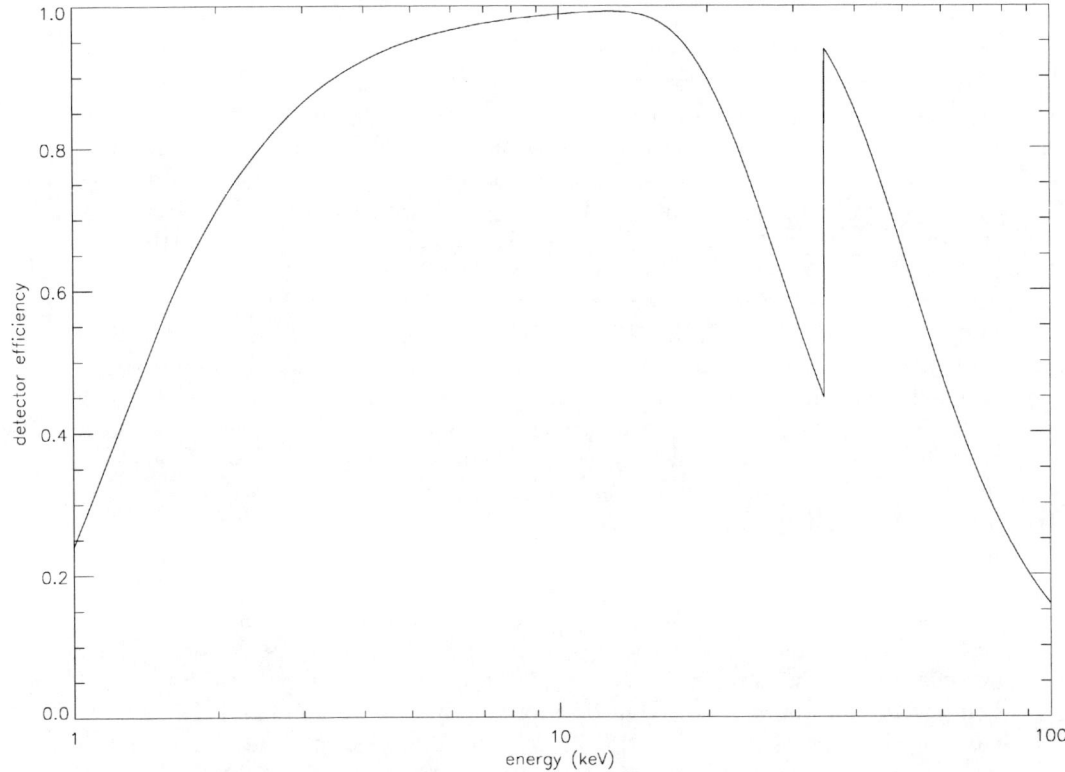

FIGURE 2. Detector efficiency as a function of photon energy.

and the energy resolution will be $\frac{\Delta E}{E} = 0.19\sqrt{\frac{5.9 \text{ keV}}{E}}$ ($E < 35$ keV), $\frac{\Delta E}{E} \leq 0.05$ ($E > 35$ keV). Special features have been implemented in the design of the X-ray detector to reduce the internal background by more than 95%, and to improve the efficiency at high energies (internal anti-coincidence and fluorescence gating). The sensitivity of the detector is shown in Fig. 3, together with the expected flux from two typical sources. It can be clearly appreciated that *SIXE*'s capabilities will allow its utilization in the study of the variability of the whole kind of proposed targets. In order to avoid the confusion of sources, while being compatible with MINISAT's pointing capability, each detector will have an hexagonal honeycomb collimator, providing an approximately circular FOV of $\sim 1 \deg \times 1 \deg$. Particular attention has been paid to the design of the detector front-end electronics and experiment data system, in order to reduce the expected dead time of the detector below ~ 25 μs, and attain a timing accuracy of ~ 2 μs. In Table 1 we have summarized the main characteristics of the X-ray detector of *SIXE*.

The SAAD will be a simple semi-conductor device aimed at determining if the fluxes of electrons and ions surpass some threshold, above which the detector high

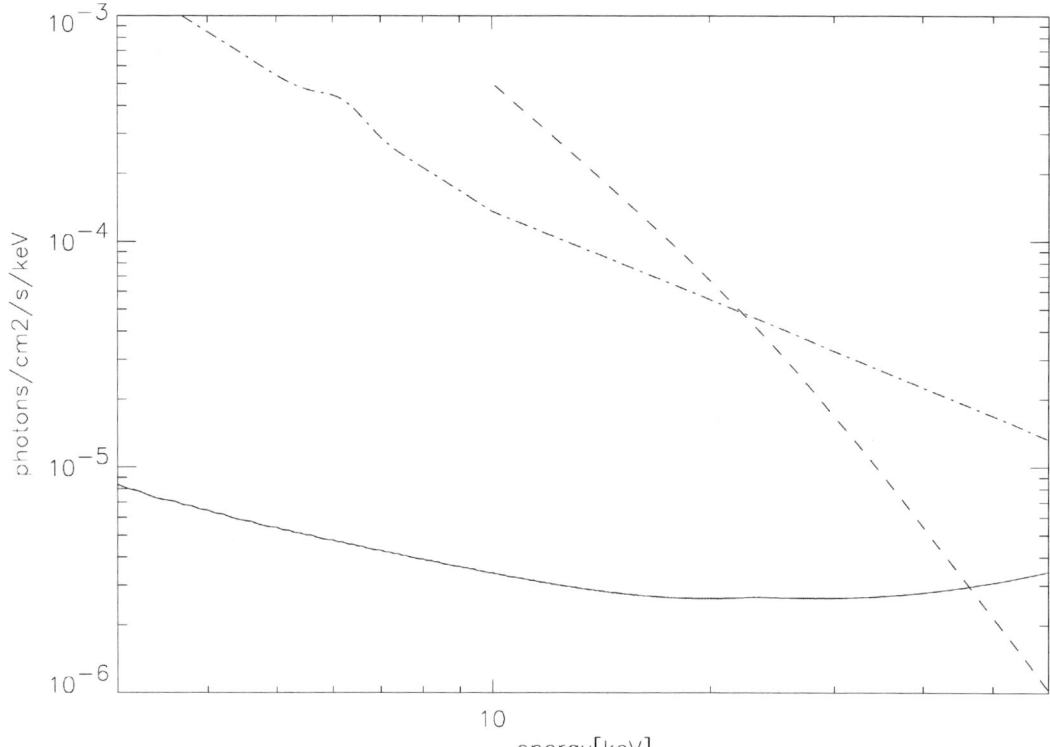

FIGURE 3. Detector sensitivity compared to the flux from two interesting extra-galactic sources. Solid line: Continuum sensitivity of *SIXE* (3σ, 10^5 s integration time. Dot-dashed line: Continuum flux from MCG-6-30-15. Dashed line: Continuum flux from the COMA cluster.

TABLE 1. Main characteristics of the X-ray detector of *SIXE*

Energy range	3-50 keV
On-source sensitive area	~ 1000 cm^2
Off-source sensitive area	~ 1000 cm^2
Integrated detection efficiency	82% (3-50 keV), 72% (3-70 keV)
Continuum sensitivity at 50 keV	3×10^{-6} ph·cm^{-2}·s^{-1}·keV^{-1} (3σ, 10^5 s)
Line sensitivity (6.7 keV, narrow line 1 keV)	10^{-5} ph·cm^{-2}·s^{-1} (3σ, 10^5 s)
Saturation rate	$\sim 50,000$ counts/s
FOV	~ 1 deg circular (FWHM)
Collimator transparency	88%
Temporal resolution	1 μs
Timing accuracy	better than 2 μs (absolute)
Spectral resolution (FWHM)	$\sim 5\%$ ($E > 35$ keV), $\sim 46/\sqrt{E}$ % ($E < 35$ keV)
Dead time	16-20 μs
Residual instrumental background	~ 63 counts/s

TABLE 2. Scientific requirements of *SIXE* compared to MINISAT-01 performance and to the maximum performances attainable with a MINISAT platform

	SIXE requirements	MINISAT-01	MINISAT maximum performances[a]
Payload mass	103.1 kg	100 kg	300 kg
Payload power consumption	58.5 W	40 W	360 W
Total data storage	2 Gb (OBC+EDS)	32 Mb	1 Gb
Data rate transmission	60 kbps (daily average)	1 Mbps	1 Mbps
Attitude stabilization	3 axes	3 axes	3 axes
Pointing error	< 0.15 deg	3 deg	< 0.1 deg

[a] The MINISAT-0 platform series is developed and operated by the INTA

voltage will be switched off. The SAAD will thus be a simple counter of events, in which the nature of the incident particle (either electrons or ions) will be deduced from the deposited energy. The data analysis and the decision of switching off the MWPC high voltage could be performed by the on board Experiment Data System (EDS).

The Optical Monitor Camera is under development by the LAEFF (INTA-CSIC)[2]. A similar device will fly on board the *INTEGRAL* mission[3]. The OMC consists on an optical system focused onto a CCD detector (1024 × 2048 pixels), with the axis co-aligned with the X-ray detector. The focal length is 153.7 mm, the aperture is 50 mm, and the (square) FOV 5 deg ×5 deg. The OMC will provide photometry in 2-3 bands, including a V band centered at 550 nm. The limiting magnitude in V is 19.7 (10×100 s, 3σ).

The overall requirements of the *SIXE* instruments are fully compatible with a small mission like MINISAT-01. A comparison of *SIXE* requirements with MINISAT-01 capabilities and with the maximum performances attainable within the MINISAT series [4] is given in Table 2.

ACKNOWLEDGMENTS

This study has been sponsored by the ESP97-1784-E grant of the Spanish PNIE (CICYT).

---

[2] http://www.laeff.esa.es/
[3] Giménez, A., et al., in *Proc. of the 2nd INTEGRAL Workshop "The transparent Universe"* (1997).
[4] From http://www.inta.es/

A Silicon-strip Coded-mask Imager for 3-40 keV X-rays

D.A. Leahy

Department of Physics and Astronomy
University of Calgary
Calgary, Alberta, Canada T2N 1N4

Abstract. An all-sky survey in the hard x-ray band (3-40 keV) at significantly higher sensitivity than the HEAO A-1 survey would be valuable in finding new sources in a number of categories, both galactic and extragalactic, for more detailed studies. A silicon-strip coded-mask imager for X-rays in this energy band has been proposed, by a team headed by the author, to carry out such an all-sky survey. Recently, the Canadian Space Agency has approved a concept study to carry out a preliminary design of this instrument.

INTRODUCTION

The 3-40 keV x-ray band includes emission from the hottest thermal objects in the universe (such as young supernova remnants and clusters of galaxies), as well as the brightest part of the spectrum from many non-thermal high energy sources (such as galactic black hole sources and active galactic nuclei). Thus observations in the 3-40 keV x-ray band are important for many studies in astrophysics. The sky has been surveyed in a similar band (2- 60 keV) by the HEAO–1 A2 instrument (for details see http://legacy.gsfc.nasa.gov/docs/journal/heao1-a2_5.html). This survey was a non–imaging survey with low sensitivity. However it did achieve major science breakthroughs, including identification of strong non–thermal emission from active galactic nuclei and measurement of the diffuse cosmic x–ray background. The HEAO–1 A2 catalogs are still used to identify sources for detailed study with pointed instruments which have much greater sensitivity. However, to find and study x-ray sources at fainter levels what is needed is a new survey with much better sensitivity. Such a survey would provide extragalactic sources at: i) much greater cosmological distances; and ii) to much higher level of completeness of sampling, for nearer sources.

An instrument for carrying out a new high-sensitivity all-sky survey is now under study, under the direction of the author, with the intention of implementation by the Canadian Space Agency. The instrument is proposed to use a coded-mask for imaging, and a silicon strip detector to provide a position sensitive detector with large area.

I SCIENCE OBJECTIVES

Here a brief summary of the types of science that would be enabled by a new high-sensitivity all-sky survey in the 3-40 keV band is given. The specific goals for the spatial and spectral resolution of the hard x-ray survey are: arcminute resolution and sub-keV spectral resolution.

Galactic sources include the following. LMXB (low mass x–ray binaries) include the brightest sources in the galaxy. These sources have been grouped into a number of categories, including the Z and atoll type sources, based on their time variability behaviour. However their behaviour is not yet understood especially in the hard x–ray band. A survey in hard x–rays is needed to identify more of these sources, and a pointed study of a few of these sources would add valuable data concerning time variability in hard x–rays.

X–ray pulsars are those x–ray binaries with a neutron star that has been detected pulsating in x–rays. Sub-categories of this class are based on the type of stellar companion to the neutron star, and include high mass x–ray binaries, Be companion x–ray binaries, and low mass x–ray binaries. The observer has a direct view of the accretion region in these systems, so the x–ray emission is a direct probe of the neutron star surface, or very near to the surface.

Only a few of the nearest and most rapidly spinning radio pulsars have been detected in x–rays, with emission of a power law spectrum of high energy radiation. The ideal way to search for more of these systems, which are key to understanding how radio pulsars work, is through a sensitive all–sky survey in hard x–ray, such as is proposed here.

Black hole x–ray sources are of two types: steady sources, such as Cygnus X–1; and transient sources, such as A0620−00. An x–ray survey of higher sensitivity than HEAO–1 would allow discovery of several more sources in the black hole category, as well as monitoring the activity of known sources.

Stellar winds from early type stars have velocities of 1000 km/s or greater, resulting in high shock temperatures (and thus in emission in the hard x-ray band). Stellar coronal sources have been found to have a range of temperatures of emission within a single object. New observations in the hard x–ray band would serve to detect new sources and study the behaviour of known sources. This is particularly important since the hard x–ray behaviour of these sources has not been adequately studied before. The main–sequence M–type stars often show significant flare activity that is analogous to the solar flares, which might be seredipitously observed during the survey. If so it would provide valuable data for comparison with solar flares. Novae are regular occurences in our galaxy and sources of hard x–rays early in their evolution. However, previously, inadequate sky coverage, sensitivity and spectral resolution have not allowed adequate study of their spectral evolution.

Extragalactic sources have been studied with pointed instruments based on their identification in the sample of hard x–ray sources published by [1], which is based on data from the HEAO–1 A2 experiment (2–10 keV). This catalog contains a total of 85 sources at galactic latitudes greater than 20 degrees, 70% of which are extragalactic. A much larger hard x–ray selected sample, which would be the result of the proposed survey, is important to obtain for the following reasons: i) AGN samples found in the hard x–ray will allow tests of the AGN unification schemes which propose obscuring tori. ii) In units of νF_ν, AGN emit the majority of their power in the hard x–ray band, and so this region is probably the driving force in AGN physics. Information on the spectral index in this region will test energy generation models such as the synchrotron self–compton idea. iii) Because of their larger redshift, only very small samples of luminous QSOs have been found in the hard x–ray and a larger sample is needed. iv) Rich clusters of galaxies have gas temperatures in the 5–15 keV range. The luminosity function and evolution of such regions places tight constraints on the fluctuation spectrum in the early universe and the fundamental cosmological parameters such as Ω and Λ. v) There is also no sizeable sample of normal galaxies observed in the hard x–ray band, as such galaxies are relatively x–ray faint. Hard x–ray information on normal galaxies will allow us to determine the incidence of x–ray binaries and SNR as a function of Hubble type and luminosity. vi) There is still considerable uncertainty about the origin of the x–ray background. The spectral shape is reasonably well fit with a bremsstrahlung shape and a temperature of 40 keV, and so luminosity functions and evolutionary models of sources in this spectral region are essential to diagnose whether additional contributors to the traditional sources (AGN and clusters) are needed.

II THE PROPOSED INSTRUMENT

The proposed instrument combines coded mask imaging in x–rays with solid-state detectors to achieve simultaneous good spatial and spectral resolution in a low-cost package. Silicon strip detectors simultaneously record the energy and position of detected x–ray photons. Silicon strip detectors were developed for particle physics detector applications, resulting in high performance yet low cost. The silicon strips for the NASA GLAST gamma–ray mission have demonstrated the low total power requirement for a large area (few thousand cm^2) silicon strip detector.

To illustrate the scientific advances that can be achieved with the new survey, it is compared the only previous hard x–ray survey, the HEAO–1 A2 survey done in 1978-1980. The HEAO–1 A2 survey had a sensitivity of $2 \times 10^{-11} erg cm^{-2} s^{-1}$ in the 2–20 keV x–ray band and detected about a hundred sources. The estimated sensitivity for the new instrument is $\sim 10^{-13} erg cm^{-2} s^{-1}$ in the 2–10 keV band. This is a dramatic improvement and will allow the new survey to detect about 5000 sources. A second dramatic improvement is that the silicon strip detectors have an energy resolution of order 0.1 keV, which is more than a tenfold improvement over

HEAO–1 A2.

The surveys can be carried out by a scanning satellite (one axis stabilized). A desirable orbital configuration to cover the whole sky and to keep low detector particle background, is low earth orbit with the spin axis of the satellite pointing at the sun and the detector axis normal to the sun. After the completing an all-sky survey, the instrument could be used to carry out a pointed phase for detailed studies of specific sources. This would require a 3–axis stabilized spacecraft with arc–minute pointing.

The current status of the proposed instrument is as follows. The Canadian Space Agency issued a call for proposals for "Concept Studies" for instruments for space astronomy at the end of 1997. The current proposal was selected in mid–1998, and the work on the concept study is starting in early 1999. The study will last several months and draws on the expertise of scientists and industry. The goal of the study is to recommend the general specifications of the instrument and spacecraft required to carry out the hard x-ray survey. Subsequently, the instrument will be proposed for detailed phase A type studies.

REFERENCES

1. Piccinotti et al., *Astrophysical Journal* **253**, 485 (1982).

Discovery Space for Stellar Astrophysics by Small Missions

Jeffrey L. Linsky

JILA/University of Colorado and NIST
Boulder Colorado 80309-0440

Abstract. The last 20 years has been a time of major advances in our ability to observe stars at a variety of wavelengths across the electromagnetic spectrum. Many questions concerning the existence of phenomena (e.g., coronae, winds, accretion disks, nonthermal particles, shocks, magnetic fields) are now answered, but important details about the phenomena and questions concerning the physical processes responsible for the observed phenomena remain unanswered. Well-conceived small space missions can answer many of these questions. I describe one small mission – Wide Angle Stellar Coronal Explorer – that could answer many of these questions at modest cost.

I INTRODUCTION

During the last 20 years, the field of stellar astronomy has benefited enormously from powerful new instruments in space and on the ground that can observe stars across the electromagnetic spectrum. It is no longer possible to list the many accomplishments that have emerged from the analysis of the excellent data from these instruments with the help of increasingly sophisticated theory, but to wet your appetite I will list a few:

- In the radio regime, the VLA ($\geq$1981) has been the prime observing instrument at centimeter wavelengths, but the Australian Telescope Compact Array, Arecibo ($\geq$1963), VLBA ($\geq$1993), several millimeter arrays, and VLBI observations with intercontinental baselines have enabled us to study mass loss from both hot and cool stars, gyrosynchrotron emission from relativistic electrons in the coronae of single and binary stars, thermal emission from the disks of young stars, and molecular emission and masers formed in the circumstellar environments of both young stars and those at the tip of the red giant branch.

- Infrared observations from the ground, the Kuiper aircraft (1975–1995), and the IRAS (1983) and ISO (1995–1998) satellites have provided critical data for studying highly obscured protostars and T Tauri stars, the very cool brown dwarfs, and the circumstellar environments of very cool giants.

- Optical observations primarily from the ground, but also from HST's very high resolution cameras, are providing high resolution images of circumstellar envelopes, measurements of stellar magnetic fields, chemical abundances, and the dynamics of gas in the lower atmospheres. Interferometry, Doppler imaging, and Zeeman Doppler imaging are now beginning to resolve stellar surfaces.

- UV spectroscopy pioneered by IUE (1978–1996) and extended by the GHRS (1990–1997) and now the STIS ($\geq$1997) instruments on HST study the 10^5 K plasma above the photospheres of stars, the dynamics and energetics of this plasma, mass loss from hot stars and cool giants, mass exchange in binary systems, accretion phenomena in premain sequence stars, and explosive phenomena in flare stars, cataclysmic variables, novae, and supernovae. Extreme ultraviolet spectroscopy was initiated by EUVE ($\geq$1992) and will be continued with much higher sensitivity by the LETG instrument on Chandra. FUSE ($\geq$1999) will observe many stars in the far ultraviolet (912–1180 Å) spectral range.

- The Einstein (1978–1981) satellite pioneered stellar x-ray astronomy and EX-OSAT (1983–1986), ROSAT (1990–1999), ASCA ($\geq$1993), and BeppoSAX ($\geq$1996) have continued these studies. Since x-rays are emitted by plasmas hotter than about 10^6 K, these satellites provide information on stellar coronae and high speed accretion flows and shocks. We have learned that shocks in the radiation-driven winds of hot stars (spectral classes O and early-B) emit x-rays, and that most other stars (except for cool giants and supergiants) have hot coronae that emit x-rays. The heating mechanisms for these hot coronae are likely magnetic in character and have a maximum heating rate at the observed saturated x-ray emission level. Classical T Tauri stars, however, often exceed this saturated emission rate likely due to an additional heating source, accretion. Very little is known about the nature of the emission processes (i.e., whether they are thermal or nonthermal) for $E > 10$ keV. We look forward to the new field of x-ray spectroscopy that will commence with the launch of Chandra (formerly called AXAF), XMM, and Astro-E.

- The highest energy emission from stars occurs during powerful flares from M dwarfs, spectroscopic binary systems, and pre-main sequence stars. The hard x-ray detector on BeppoSAX has seen a few stellar flares and the Compton Gamma Ray Observatory ($\geq$1991)and solar instruments have detected MEV emission lines and continuum from solar flares.

II SOME MAJOR UNSOLVED PROBLEMS IN STELLAR ASTROPHYSICS

Thanks to these successful missions, we have been able to study at some level essentially all types of stars of all ages across the electromagnetic spectrum. These observations have provided insights into the phenomenology of stars including their formation and evolution, mass loss, high energy phenomena (coronae and flares), binary interactions, circumstellar phenomena, magnetic fields and cycles, angular momentum evolution, internal dynamics and evolution, and nonradiative heating processes. Nevertheless, new data inevitably lead to more refined (and hopefully deeper) questions about the physical processes that underlie the observed phenomena. These questions can form the starting point in planning for future missions. Since this meeting is oriented to possible future missions, I list a few important but unanswered questions about stars that should be addressed in future missions. This list includes those questions of particular interest to me, but the list is far from exhaustive:

Why do stellar coronae not show cycles? The solar magnetic field reverses polarity with a 22 year cycle. Twice during the cycle the number of sunspots increases dramatically, flares and high energy phenomena reach a maximum, and the x-ray, UV, and radio emission go from quiescent levels at solar minimum to relatively high stars at solar maximum. During the solar 11 year magnetic cycle, the ratio of maximum to minimum x-ray luminosities, $L_X^{max}/L_X^{min} \approx 10$ in soft x-rays and ≈ 200 in hard x-rays. Many stars should show cycles analogous to the solar cycle. Do they?

- Magnetic field strengths are now measured in many late-type stars using Zeeman broadening of unpolarized absorption lines. For a few stars Zeeman Doppler images are also now available. There are, however, no magnetic field measurements to date that directly show that a star has a magnetic cycle. It is possible that the lack of direct detections of magnetic cycles is a consequence of too short a time baseline for observations and no long term systematic program to monitor stellar magnetic fields on a few interesting stars, but I am surprised by the absence of even one directly detected magnetic cycle.

- Bright emission in the cores of the Ca II resonance lines (the so-called H and K lines) is known to be a strong indicator of nonradiative heating in the solar chromosphere and is a good proxy of magnetic field strength on the Sun. Since 1966 monitoring of Ca II emission in late-type stars from Mt. Wilson has discovered periodic variations in the Ca II emission in a number of stars, typically main sequence stars that are middle-aged and relatively slow rotators [1]. Thus magnetic cycles must be present in many stars.

- There is as yet no conclusive evidence for periodic coronal x-ray variations on stellar cycle timescales on any star except the Sun. For example, Einstein and ROSAT observations of the Hyades cluster stars (age 630 Gyr) obtained 10 years apart show less than a factor of 2 flux differences. Observations of the Pleiades cluster stars (age 60 Gyr) show more variations, but no cyclic behavior? Other clusters are listed in Table 1. Is the Sun unique (unlikely given the Ca II stellar monitoring data), or do we have too few data to make an intelligent statement? This is a question that begs for a future observing program.

TABLE 1. X-ray observations of clusters.

Cluster name	Diameter (Degrees)	log(age)	Distance (parsecs)	Number of x-ray sources	Reference
Orion	4	6.	460	389	ApJ 445, 280 (1995)
Chamaeleon I	2	6.8	140	89	ApJ 416, 623 (1993)
Rho Oph	2	6.8	200	55	ApJ 439, 752 (1995)
Pleiades	2	7.8	116	85	ApJ 348, 557 (1990)
Coma	5	8.6	88	102	A+A 313, 815 (1996)
Hyades	40.	8.8	46	108	ApJ, 399, L159 (1992)
Praesepe	1.5	8.9	177	68	A+A 298, 115 (1995)

How and why are the outer atmospheres of "active" stars different from "inactive" stars? Photospheric magnetic field strengths tend to be close to their equipartition values ($B_{eq}^2/8\pi \approx P_{gas}$), but the fraction of the photosphere covered by strong magnetic fields, the magnetic filling factor, increases with rotation velocity from about 1% (Sun) to > 50% for active stars. This has important consequences. For example, in an atmosphere with a small magnetic filling factor, the field lines can expand rapidly with height, whereas in an atmosphere with a large magnetic filling factor the field lines can expand only very slowly. As Cuntz and Ulmschneider [2] have pointed out, upwardly propagating MHD waves shock low in the chromosphere and thus rapidly heat the plasma in a flux tube that does not diverge rapidly with height, whereas MHD waves may not form shocks at all in a flux tube whose cross-sectional area increases rapidly with height. This is one example of how the magnetic atmospheres in "active" stars may differ from the nonmagnetic atmospheres of "inactive" stars. There is much more to be learned concerning this topic.

What is the flare energy budget for young and old stars? There is a general trend in which the normalized coronal heating rate (measured by L_X/L_{bol}) increases with stellar rotation rate (and decreasing age) to reach a maximum value (called "saturation") at 10^{-3}. One typically plots the L_X/L_{bol} ratio with respect to the Rossby number, $N_R = P/\tau_c$, where P is the rotation period and τ_c is the convective turnover time. The Rossby number characterizes the properties of $\alpha\Omega$-type dynamos. Two violations of this general trend need to be addressed:

- Why does L_X/L_{bol} decrease for very rapid rotation rates? This unexpected phenomenon is often called "supersaturation"?

- Why for Classical T Tauri stars does L_X/L_{bol} often exceed the 10^{-3} saturation limit? Is this because accretion provides an additional energy source above the usual magnetic heating processes?

Are stellar coronae heated by flares and microflares? Rapid increases in the high energy emission, generally called "flaring," is usually thought to be caused by the rapid conversion of magnetic energy to heat, relativistic particles, and explosive expansion of the plasma. A long record of stellar x-ray emission of an "active" star would identify the distribution of flare energies, leading to an extrapolation to smaller (undetected) flare energies and the total flare energy budget. Unfortunately this experiment has not yet been done because it is difficult to acquire a very large block of observing time to study such stars on large satellites, which have competing requests for observing time. A small satellite, however, can devote the time to such a long monitoring program.

What is the relation between thermal and nonthermal phenomena in stellar coronae? Soft x-ray emission (0.1 – 8 keV) is usually explained by thermal bremsstrahlung from a hot plasma, but centimeter-wave radio emission is modelled as gyrosynchrotron emission from nonthermal electrons on the basis of the high brightness temperatures, negative spectral indicies, polarization, and flux at high radio frequencies.

- What are the physical and phenomenological relations of the thermal and nonthermal plasmas? Are they entirely separate plasmas? Is one electron energy distribution the source of the other?

- Very high temperature plasma (or nonthermal plasma) is seen during flares For example, the analysis of a BeppoSAX observation of the flare star AD Leo [3] shows a very hot component that if thermal has kT > 18 keV. Is such hot (or nonthermal) plasma a common occurance in flares? If so, the nonthermal electrons that emit radio radiation during flares may be the high energy component of the electrons observed at x-ray wavelengths.

- Analysis of solar hard x-ray emission typically leads to models in which the thermal plasma has a nonmaxwellian power law tail at high energies. Can one detected such nonthermal tails in the 10–20 keV energy range in stars?

What physical processes are occuring in the winds of hot stars? The standard model of hot star winds describes winds that are accelerated by radiation pressure. Since this acceleration process is unstable, shocks occur in the wind that produce hot plasma observed in the x-ray regime.

- Is the model of x-ray emission from shocks in radiatively driven winds basically correct, or has some important process been left out?

- What is the physical explanation for the observed relation $L_X/L_{bol} \approx 10^{-7}$ for these stars?

- Is the x-ray emission produced in many small shocks or in a few large shocks?

What physical processes occur in the winds and coronae of cool stars?
The coolest giants and supergiants have low-temperature winds with high mass loss rates but no x-ray emission, whereas main sequence stars with spectral types F–M and giants with spectral types G and early-K have small mass loss rates and hot coronae? What physical processes lead to this bifurcation and the intermediate case of the hybrid-chromosphere stars which have strong winds and x-ray emission?

- Some Pleiades-aged stars like AB Dor show evidence for very large ($> 3R_*$) loops that corotate with the star out to at least the Keplerian radius. These loops are also seen in the Hα line. Do such large features have coronal counterparts?

- What is the spatial relation of starspots, presumably regions of the photosphere with the strongest magnetic fields, with coronal active regions?

- How spatially extended are stellar coronae? Observations of binary systems as one star eclipses the other could answer this question.

- What is the temporal relation of x-ray and microwave flares? On the Sun x-ray emission from the hot thermal plasma peaks about five minutes after the initial hard x-ray spike and initial nonthermal radio emission. Is the same true for different classes of flaring stars?

- Do all brown dwarfs have coronae? The very coolest M dwarfs are x-ray sources, and there are now ROSAT observations of x-ray emission from a few brown dwarfs in young clusters [4]. Do the older brown dwarfs in the field have coronae? How are these coronae heated?

III   DISCOVERY SPACE AND SMALL MISSIONS

Progress in our understanding of stellar astrophysics requires pushing the envelope of observational capability beyond present standards. This does not necessarily require large, expensive space missions. Suitably designed small missions with appropriate operational philosophies have important roles to play. Table 2 lists some important measurement objectives needed to address the questions listed above and other important questions not listed. Table 2 also lists which, if any, of the large missions are well designed to meet these objectives and which measurement objectives are best accomplished by small missions.

IV   A PROPOSED MISSION – WIDE ANGLE STELLAR CORONAL EXPLORER

Science Objectives of the Mission

TABLE 2. Missions that can achieve different measurement objectives.

Measurement Objective	Mission size that is most suitable for each objective	
	Large Missions	Small Missions
Higher energy resolution	Chandra, XMM, Const-X	No
Greater x-ray sensitivity	Chandra, XMM, Const-X	No
Greater UV sensitivity	HST/COS	No
Higher time resolution (x-ray)	Chandra, XMM, Const-X, XTE	No
Higher time resolution (UV)	HST	No
Rapid access to TOOs (x-ray)	No	Yes
Rapid access to TOOs (UV)	No	Yes
Simultaneous x-ray and EUV	Chandra	Yes
Simultaneous UV and x-ray	XMM	Yes
Simultaneous soft and hard x-ray	Const-X, SAX	?
Wide angle x-ray imaging	No	Yes
Wide angle UV imaging	No	Yes
Long term x-ray monitoring	No	Yes
Long term UV monitoring	No	Yes

Stellar x-ray cycles: Search for periodic x-ray emission on time scales of years that would indicate cycles in stars with different ages, spectral types, and rotational velocities. This objective requires a long-lived satellite (5–10 years).

Rotation: Search for periodic x-ray emission on time scales of days that would indicate rotation periods. These data, when combined with optical measurements of *vsini* and estimates of radius, provide a measure of the inclination of the rotation axis to the line of sight.

Supersaturation: Why does L_x/L_{bol} decrease with increasing *vsini* for very active stars? Could this be due to declining flare activity in the most rapidly rotating stars?

Do flares heat coronae? Long term monitoring of the x-ray emission from stars will measure the fraction of the x-ray emission that is flare related for different types of stars. These data will also measure the flare energy distribution that can be extrapolated to lower energies to estimate the total contribution of flares to the x-ray emission.

Coronae of O stars: Is the x-ray emission from these stars produced in a large number of shocks in their radiatively-driven winds, or do a few (perhaps magnetic) shocks dominate the emission which is time variable?

Coronae of A and B-type stars: What is the temporal behavior of the x-ray emission from young A and B-type stars? Is there a steady component with flares or only flare emission? What causes some of these stars to be x-ray emitters?

Coronae of T Tauri stars: For low-mass stars with $L_x/L_{bol} \geq 10^{-4}$, is the x-ray emission produced in an enhanced corona and/or from the interaction of magnetic fields in the stellar corona and disk as proposed by Shu et al. [5]?

Brown dwarfs: Look for x-ray flares from substellar mass ($M \leq 0.07 M_\odot$) brown dwarfs.

Mission requirements

Field of view: At least 3 degrees radius at one time.

Energy range: Either broad band soft x-rays for grazing incidence optics, or higher effective area but limited passband for normal incidence multilayer optics.

Angular resolution: About 10 arcseconds.

Energy resolution: Similar to ACIS.

Effective area: As large as feasible but at least 300 cm^2.

UV passband: It is desirable but not required to have UV imaging capability with a 1300–1600 Å passband.

Mission duration: Minimum of 5 years with a goal of 10 years.

Time resolution: Better than 1 second.

Orbit: HEO highly desirable to permit long duration observation of targets with minimal interruption by Earth occultation.

Observing strategy: Monitor the x-ray emission from a young cluster continuously for a month and then return after several months on a regular basis. Select 3 or 4 clusters with different ages for this program.

Targets of opportunity: Ability to slew and start observing a target of opportunity within 5 minutes of a decision (which could be automatic based on preset criteria).

REFERENCES

1. Baliunas et al., *Ap. J.* **438**, 269 (1995).
2. Cuntz M., Rammacher W., Ulmschneider P., Musielak, Z.E., and Saar S.H., *Ap. J.*, in press (1999).
3. Sciortino S., Maggio A. Favata F., and Orlando S., *Astron. Astrophys.* **342**. 502 (1999).
4. Neuhauser R., et al. *Astron. Astrophys.* **343**, 883 (1999).
5. Shu F., Shang H., Glassgold A.E., and Lee T., *Science* **277**, 1475 (1997).

Instrumentation for a Next-Generation X-Ray All-Sky Monitor

A. G. Peele

Code 662, Laboratory for High Energy Astrophysics, Goddard Space Flight Center, Greenbelt, MD 20771

Abstract. We have proposed an x-ray all-sky monitor for a small satellite mission that will be ten times more sensitive than past monitors and that opens up a new band of the soft x-ray spectrum (0.1 - 3.0 keV) for study. We discuss three approaches to the construction of the optics. The first method, well within the reach of existing technology, is to approximate the lobster-eye geometry by building crossed arrays of planar reflectors, this gives great control over the reflecting surface but is limited in terms of resolution at the baseline 4 arc minute level. The second method is to use microchannel plates; this technology has the potential to greatly exceed the baseline resolution and sensitivity but is yet to be fully demonstrated. The third method, while still in its infancy, may yet prove to be the most powerful; this approach relies on photolithography to expose a substrate that can then be developed and replicated.

The scientific case for this mission is almost too broad to state here. The instrument we describe will allow investigation of the long term light curves of thousands of AGN, it will detect thousands of transients, including GRBs and type II supernova, and the stellar coronae of hundreds of the brightest x-ray stars can be monitored. In addition the classical objectives of all-sky monitors — long-term all-sky archive and watchdog alert to new events — will be fulfilled at an unprecedented level. We also note that by opening up a little-explored band of the x-ray sky the opportunity for new discovery is presented. A satisfying example of entering new territory while still retaining the guarantee of expanding the domain of existing research.

INTRODUCTION

All-sky monitors ('ASMs') have traditionally played a central role in x-ray astronomy. Their wide field of view has allowed ASMs to monitor many bright sources, as well as discovering large numbers of transients. However, previous ASMs were limited in sensitivity to x-rays above 2 keV and had relatively small effective areas. Additionally, they have not possessed focusing capability. As a result their sensitivity has at best been of order several mCrab in a few hours. Typically, the strength of ASMs has been in monitoring galactic x-ray binaries. In order to monitor other sources such as active galactic nuclei, coronal sources, cataclysmic variables and white dwarf binaries soft x-ray sensitivity of the order of a few tenths of a mCrab is required. In order to reach this limit with a non-focusing device the instrumental mass would have to be of order 10^4 kilograms (1). To attain high sensitivity, while remaining within reasonable budgetary limits, optics which offer both focusing and a wide field of view are desirable.

The concept of using lobster-eye optics (2), or their one-dimensional counterpart (Schmidt design) (3), has long been of interest in x-ray astronomy. These systems offer compact focusing and wide field of view. However, the difficulty of manufacturing optics with the required tolerances has held this field in abeyance. Recently, developments in the manufacture of microchannel plates (MCPs) (4, 5, 6) have suggested that the technology has advanced sufficiently to construct a working lobster-eye optic. Similarly, it is now also possible to construct fixtures with less than one arc-minute alignment errors, which means a sufficiently small one-dimensional system can be made.

We have previously considered the performance of an optimal two-dimensional lobster-eye system (7) based on MCPs. Here we also examine the sensitivity and modular construction for a one-dimensional system based on a modification to the Schmidt design. In addition, we will consider a two-dimensional system with parameters that might be obtained using photolithographic methods.

PRINCIPLE OF OPERATION

Named for the crustaceans that evolved a similar optical system (8), the lobster-eye optic is a square packed array of square channels. The long axes of the channels intersect at a distance R from the center of the channels (Fig. 1). The optic brings grazing incidence x-rays to a focus at a distance $R/2$ behind the array. Photons form a point focus if they reflect an odd number of times in each orthogonal axis. A line focus results from photons that reflect an odd number of times in one axis and an even number (which includes zero) of times in the orthogonal axis. Photons reflected even numbers of times in both axes are unfocused and form a "diffuse" background to the image (Fig. 2). The net image is cruciform with the central focus dominating for an optimized system (Fig. 3). The one-dimensional system simply consists of flat reflectors radially aligned about a center of curvature with cylindrical symmetry. Fig 1 represents the view along the cylindrical axis. Such a system on its own would result in a single line focus. It has been shown that such single systems are less sensitive, for typical parameters, than two-dimensional lobster-eye systems for all but the brightest astronomical sources (9). However, it is possible to cross two one-dimensional arrays to approximate a lobster-eye array (Fig. 4).

TECHNOLOGY

Microchannel-plates

State of the art MCP manufacture produces MCPs that focus with a resolution of approximately 3 arc-minutes (5, 6). The theoretical resolution can be much less than this. The causes for the degradation in resolution are ascribed to mechanical defects in the MCP. For instance, random long axis channel tilts with a root mean square (rms) value of 0.2 mrad has been measured in one sample (6). Also rms channel rotations of ~ 15 mrad and surface roughness rms of ~2 nm have been measured (10, 5, 11, 6).

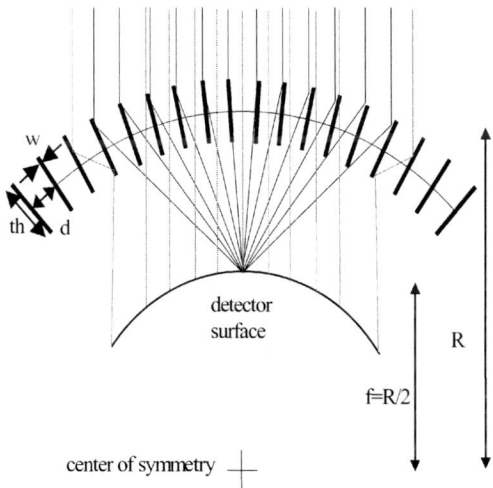

FIGURE 1: Cross section of lobster-eye optic, or view down the cylindrical axis of a one-dimensional optic.

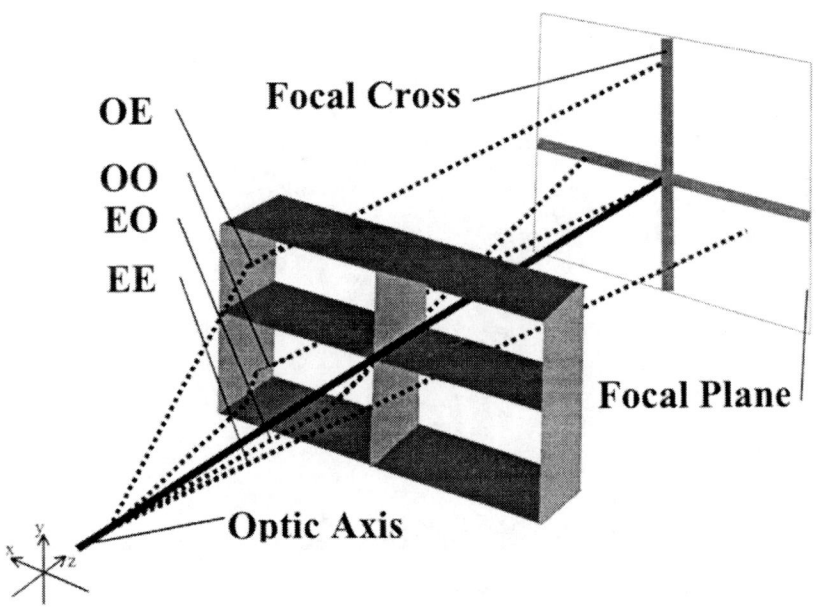

FIGURE 2: Principle of lobster-eye focusing.

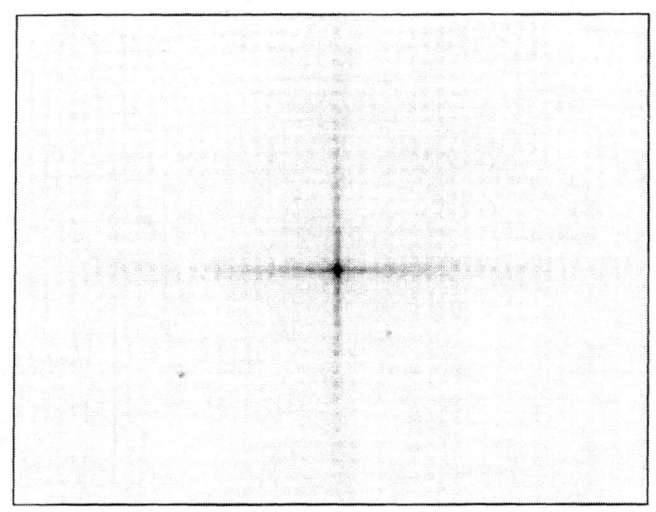

FIGURE 3: lobster-eye focusing.

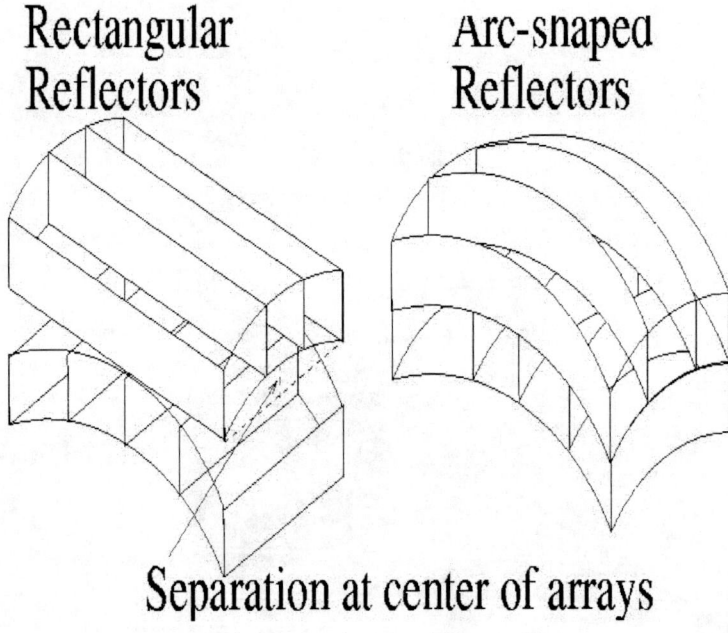

FIGURE 4: Crossed one-dimensional arrays in Schmidt format on the left and Modified Schmidt format on the right.

In addition MCPs can be susceptible to gross defects in manufacture such as a coherent variation in the channel long axis alignment caused by roller misalignment or roughness during the drawing process (a step in MCP manufacture is the repeated drawing of glass fiber bundles – see Fig.5).

This can cause distortion of the focal arms and central focus as shown in Figs 8 and 6. Another issue for MCPs is whether it is possible to coat the insides of the glass channels in order to enhance their x-ray reflectivity. A demonstration of the enhancement in reflectivity by Nickel coating an MCP is shown in Fig. 7.

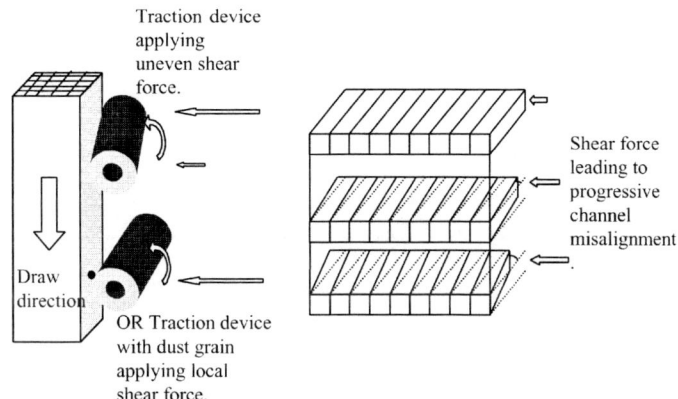

FIGURE 5: Schematics showing how an uneven shear force might be applied to the sides of an MCP during the drawing process.

Photolithography

A range of different techniques falls under this heading. For instance we have recently investigated a sample, shown in Fig. 9, that was made by a process known as electro-chemical etching (13). In this case it was shown that the sample did not possess the high geometric tolerances required to operate as a lobster-eye optic. A more promising approach is deep x-ray lithography. In this approach synchrotron radiation is exposed onto a masked substrate. The high energy and parallel nature of the synchrotron flux allows us to obtain the high aspect ratio needed for lobster-eye optics while retaining control over the geometry of the sample. The exposed substrate is then developed leaving a negative cast that is replicated in metal to obtain the final structure. It is this control over the final material of the sample that makes this approach exciting, as no coating would be required.

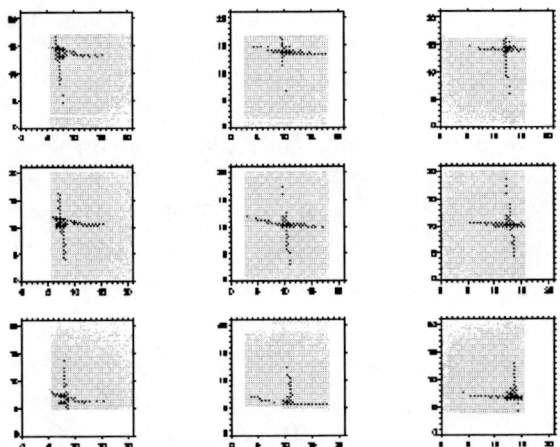

FIGURE 6: Modeled focal images, incorporating the defect described in Fig. 5 for on axis source (center) and off-axis positions.

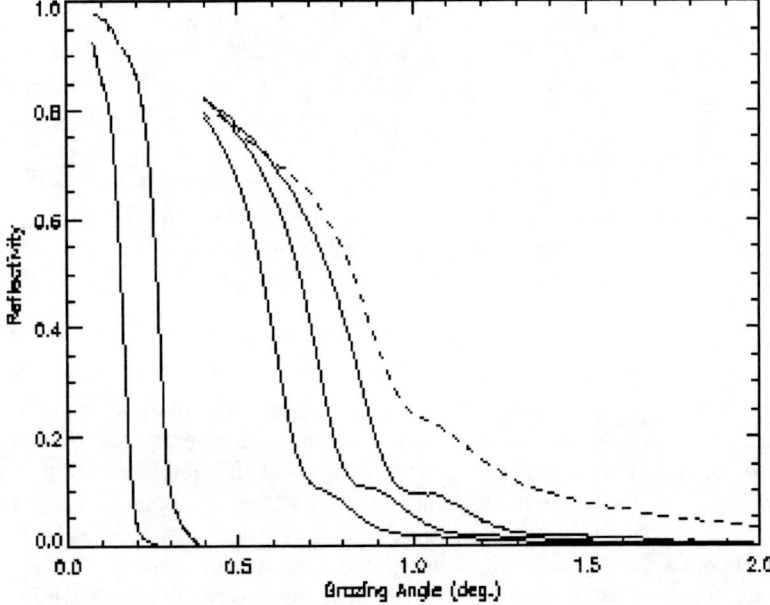

FIGURE 7: Reflectivity vs. Grazing Angle (deg.) curves for a bare MCP at energies of (from left to right) 12, 8, 2.8, 2.3, and 2.0 keV. The dashed line is for a Ni-coated MCP at 2.0 keV.

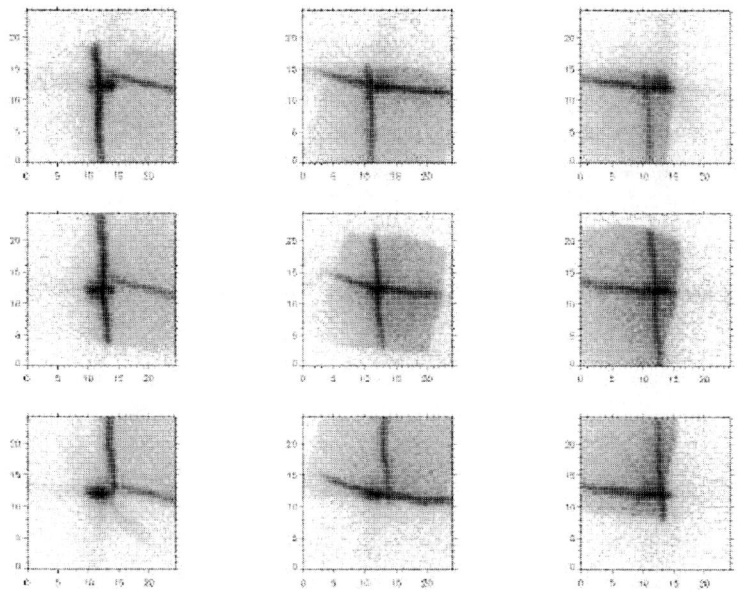

FIGURE 8: Focus obtained with source aligned on axis (center image) and off-axis along the x and y axes and along the diagonals.

FIGURE 9: Electro-chemically etched Silicon sample. Channel opening is ~ 6 μm.

Modified Schmidt Design

Ideally we would like an ASM with 4π steradians field of view. However, for the crossed one-dimensional system this is not feasible, as the second array must be moved towards the detector so as not to interfere with the first array. To accommodate this displacement the radius of curvature of the second array is adjusted so that the focus for both arrays is coincident. However, at 120° field of view the second array reaches the detector surface. In addition, before this limit is reached, the radius of curvature of the second array will have shortened so much that the angular resolution at the focus has become unacceptable. In order to avoid this problem we propose shaping the reflectors so that the interface of the array is spherical rather than cylindrical (Fig 4). This approach maximizes the radius of curvature of the second array, and hence the potential angular resolution, while retaining the same amount of reflecting material.

Even with spherical curvature of the arrays, there will be a practical limit to the field of view obtainable. As we go away from the on-axis direction the projected channel shape through the two arrays becomes non-square. This will cause the central focus arms in the image to flare out.

Another reason for limiting the field of view of a given module is the detector. The focal surface for the two-dimensional array is a sphere, while for the crossed one-dimensional arrays the focal surface consists of two cylinders. By curving shaping the arrays as in Fig. 4, the focal surface is modified slightly from two crossed cylinders, however, off-axis the image is still astigmatic. If we limit the field of view then focal blur can be kept to an acceptable level.

The advantage of the crossed one-dimensional system is that the reflectors can be coated easily. The technology limitations will be budget driven. We know from work with foil telescopes that alignment of sections is possible to well below one arc-minute. Thus we can expect to be able to align the individual reflectors to the one-arc-minute level. Better may be possible by spending more money. A related issue is the flatness of the glass. We have measured sample glass sheets on scale lengths similar to the size of the proposed reflectors and found that flatness of ~ 1 arc-minute is typical. Again it may well be possible to improve this figure by exploring more expensive options.

TABLE 1: Optics parameters.

Parameter	MCP	Photolithography	crossed one-dimensional
Radius of curvature	1.5 m	1.5 m	1.5 m
Open area	51%	51%	51%
Channel size	10 µm	10 µm	350 µm
Channel thickness	0.6 mm	0.4 mm	1.3 cm
Material	lead glass	Nickel	Nickel coated glass
Roughness	2 nm	2 nm	0.3 nm
Tilt error	0.2 mrad	0.2 mrad	0.3 mrad
Module field of view	11.5° x 11.5°	11.5° x 11.5°	11.5° x 11.5°

PERFORMANCE

TABLE 2: Performance parameters at 1.5 keV for "perfect" systems.

System	HPD (arc min)			Arm Eff. Area (cm^2)			Central Eff. Area (cm^2)		
	0°	2.5°	5°	0°	2.5°	5°	0°	2.5°	5°
MCP	0.09			8.8			5.2		
Photolith	1.0			17.3			8.6		
one-d	1.6	1.5	1.5	18.0	18.0	9.4	8.7	8.4	3.3

We examine below the imaging performance and effective area for the three different technology approaches for the parameters given in Table 1. The following features should be noted:

- The MCP parameters are realistic estimates based on existing MCP studies.
- Modules made from MCPs or by photolithography have fields of view that are, in principle, not constrained to 11.5° x 11.5° but may be as much as 4π steradians.
- The parameters for modules made by photolithography are estimates of what seems feasible based on current understanding of the technology.
- The channel thickness is, in each case, the optimised value at 1.5 keV.
- The larger channel size for the crossed one-dimensional system is set by the fixed open area and the thickness of the glass reflectors (here 140 μm).
- Surface roughness is greater for both two-dimensional cases than for the one-

TABLE 3: Performance parameters at 1.5 keV for systems with realistic defect parameters.

System	HPD (arc min)			Arm Eff. Area (cm^2)			Central Eff. Area (cm^2)		
	0°	2.5°	5°	0°	2.5°	5°	0°	2.5°	5°
MCP	2.9			7.0			3.4		
Photolith	3.1			10.8			3.9		
one-d	4.7	4.3	4.6	17.6	17.7	9.1	8.8	9.0	3.5

dimensional case where smooth float glass can be used as well as polishing.
- The other major difference between the systems is the choice of reflecting material available to each.
- The geometry chosen for the two-dimensional systems is the same "crossed-meridional" geometry as used for the crossed one-dimensional system. The difference being that orthogonal walls occupy the same volume in the two-dimensional cases, while they are displaced for the one-dimensional case. It is not clear that this type of geometry is in fact that which a slumped MCP or photolithographic array adopts. However, any error will be small as MCP tile size is typically ~ 2 cm. For photolithographic samples more care may have to be taken in defining the geometry as larger tile sizes are possible.

In Table 2 we show performance for a system that is defined only by its geometry and the reflecting material. The "defect" parameters, tilt and roughness, are set to zero. We have not calculated off-axis performance for the two-dimensional cases as, ideally, only vignetting will be a factor. In practice, alignment accuracy of the tiles, the actual channel geometry within the tiles, the size of each tile and the tiling geometry will also have an effect. The two Nickel coated systems have similar effective areas, while the effective areas for the lead-glass system are reduced as expected. The on-axis result for the photolithographic sample was obtained using a tile size of ~ 6 cm and the reduction in resolution, which is due the meridional geometry, compared to the 3 cm tile size used for the MCP case can be observed. The further reduction in resolution for the one-dimensional case is due to the effects of significant thickness in the parallel sided reflectors (8), the separation of the two arrays, and the astigmatism in the system. However, these effects can cancel out to some extent at some off-axis angles so that the on-axis position does not necessarily have the best resolution.

Table 3 shows the performance parameters when channel tilts and surface roughness are included. The dominant effect for the resolution in each system now becomes the channel tilt error, while the additional smoothness possible for the one-dimensional reflectors gives that system some advantage in effective area.

DISCUSSION

For observations of a background limited source the signal to noise ratio (SN) is:

$$SN \propto \frac{EA}{\sqrt{Ef\, A_{focus}}},$$

where EA is the effective area, Ef is the background throughput, which is roughly proportional to the effective area, and A_{focus} is the area of the central focus, which is proportional to the square of the half power diameter (HPD). So:

$$SN \propto \frac{\sqrt{EA}}{HPD}.$$

For the on-axis direction in Table 3 this figure of merit is ~ 0.64 in each of the three cases. The equivalent number for the lobster-eye telescope study described in the study by Priedhorsky, Peele and Nugent (7) is 0.57, which suggests that the systems here would have a similar sensitivity of ~2 x 10^{-12} erg cm^{-2} s^{-1} or 0.1 mCrab. With our present estimates and technology level each case is currently equally viable. However, it is fair to say that the estimates for the photolithography approach and, to a lesser extent, the one-dimensional approach are approximate. Further experimental work is required before the calculations described above can be made with more precision and a leading candidate for the construction of a lobster-eye telescope identified.

REFERENCES

(1) S. S. Holt and W. C. Priedhorsky, "All-sky monitors for x-ray astronomy," Space Sci. Rev., **45**, 269-289 (1987).
(2) J. R. P. Angel, "Lobster eyes as X-ray telescopes", Astrophys. J. **233**, 364-373 (1979).
(3) W. K. H. Schmidt, "A proposed x-ray focusing device with wide field of view for use in x-ray astronomy," Nucl. Instrum. and Meth. **127**, 285-292 (1975).
(4) H. N. Chapman, K. A. Nugent, and S. W. Wilkins, "X-ray focusing using square-channel capillary arrays," Rev. Sci. Instrum. **62**, 1542-1561 (1991).
(5) A. G. Peele, K. A. Nugent, A. V. Rode, K. Gabel, M. C. Richardson, R. Strack, W. Siegmund, "Towards x-ray focusing using lobster-eye optics: a comparison of theory with experiment", Appl. Opt. **35**, 4420-4425 (1996).
(6) A. Peele, G. Fraser, A. Brunton, A. Martin, R. Rideout, N. White, R. Petre, and B. Feller, "Recent studies of lobster-eye optics," Proc. SPIE **3334**, 404-415 (1998).
(7) W. C. Priedhorsky, A. G. Peele, and K. A. Nugent, "An X-ray all-sky monitor with extraordinary sensitivity", MNRAS **279**, 733-750 (1996).
(8) A. G. Peele and W. Zhang, "Lobster-eye all-sky monitors: A comparison of one- and two-dimensional designs", Rev. Sci. Instrum. **69**, 2785-2793 (1998).
(9) M. F. Land, "Animal eyes with mirror optics", Scientific American **239**, 88-99 (1978).
(10) H. N. Chapman, A. Rode, K. A. Nugent and S. W. Wilkins, "X-ray focusing using cylindrical-channel capillary arrays. II Experiments", Appl. Opt. **32**, 6333-6340 (1993).
(11) A. N. Brunton, G. W. Fraser, J. E. Lees and I. C. Turcu, "Metrology and modeling of microchannel plate x-ray optics", Appl. Opt. **36**, 5461-5470 (1997).
(12) A. G. Peele "Investigation of etched Silicon wafers for lobster-eye optics," Rev. Sci. Instrum., **70**, 1268-1273 (1999).

Supernova Remnant

Hiroshi Tsunemi

Graduate School of Science, Department of Space and Earth Science, Osaka University,Toyonaka, Osaka, 5600043, Japan
CREST, Japan Science and Technology Corporation (JST)

Abstract. We present here some of the observational results of supernova remnant (SNR) obtained with the *ASCA* satellite. In particular, we will focus here on the relatively young SNRs and their related topics. The *ASCA* SIS has good energy resolving power which reveals the nature of the X-ray spectra. There are two types of spectra: a thin thermal emission and a synchrotron emission. Their X-ray spectra give us a good diagnostics of the plasma. Generally, they can be fitted by using model spectra not in collisional ionization equilibrium (CIE) condition but in non-equilibrium ionization condition (NEI). Some of them show gradients both in the temperature and in the abundance. The X-ray spectrum of young SNRs, like the Kepler's SNR, can be well fitted by the superposition of the fore shock component and the reverse shock component, both of which show gradients in various parameters: temperature, density, ionization parameter etc. Cassiopeia-A, another typical young SNR, shows not point symmetric but axial symmetric both in the intensity profile and in the line center energy profile. This suggests the Doppler motion of the plasma about several thousands km/sec. SN1006 is an enigmatic young SNR. *ASCA* discovered that the shell region produced a synchrotron spectrum while the interior produced a thermal emission. This can be understood as a possible site of the high energy cosmic rays. The *ASCA* GIS has good timing resolution with moderate spectral resolution. It showed a high performance of the pulsar detection. We found that several young SNRs were associated with X-ray pulsars in them.

I INTRODUCTION

The supernova (SN) explosion is a source of heavy elements in the galaxy. After the explosion, they are left as supernova remnant (SNR) which are bright in various wavelengths. There are 215 radio SNRs [4] and 102 X-ray SNRs [2]. The X-ray spectrum of the SNR is useful to perform the diagnostic of the plasma which contains heavy elements. The young SNRs, like Cassiopeia-A [5], Tycho [6] and Kepler [8], show various emission lines in their spectra. They are mainly originated from the ejecta rather than the interstellar matter (ISM). In the young SNRs, the plasma does not reach the CIE condition. Therefore, there are three types of temperatures: the ion temperature, T_i, the ionization temperature, T_{ion} and the electron temperature, T_e. Roughly speaking, the continuum of the X-ray spectrum determines T_e and the intensity ratio of the emission lines determines T_{ion}. T_i can be determined by the Doppler broadening of the emission lines which requires much better energy resolving power as well as the atomic process in detail.

The *ASCA* GIS discovered several X-ray pulsars which might be associated with SNRs. The high time resolution, the wide energy band and the moderate spatial resolution enable us to detect the pulsation hidden in the SNR nebulosity.

II X-RAY SPECTRA FROM YOUNG SNR

A Kepler's SNR

We observed the Kepler's SNR by the *ASCA* SIS in the PV phase program. Due to the background condition and the CCD working condition, the total effective exposure time is about 15 ksec [8]. The image shows circular shape forming a shell structure. Whereas, the diameter is about 3′ which makes difficult to investigate the spatially resolved analysis by using the *ASCA* satellite. Therefore we derived the X-ray spectrum from the entire SNR.

Following to the standard background subtraction method, we obtained the spectrum with the SIS shown in figure 1. There are several emission lines clearly resolved. They are Si, S, A, Ca and Fe. There are three interesting features. One is that there is a very strong emission around 1 keV probably originating from the Fe-L line blends. The second one is that the line center energy of Fe $K\alpha$ is 6.475 ± 0.015 keV: suggesting that Fe is in very low ionization stage. This means that T_e must be very low if it is in the CIE condition. The third one is that the continuum spectrum extends to high energy, up to 10 keV suggesting that T_e must be very high.

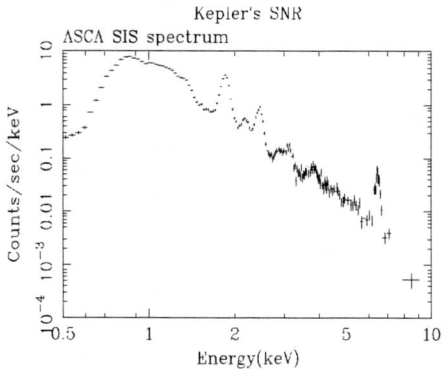

FIGURE 1. X-ray spectrum of the Kepler's SNR observed with the *ASCA* SIS.

We fitted the data by using the NEI model [12] with a single T_e component.

The NEI model we used contains an ionization parameter, τ. τ is the product of the electron density (cm^{-3}) and the elapsed time (sec) after the shock heating.

We found that two NEI components with different both in kT_e and in τ could reproduce the data well. One component shows $\log\tau = 11.04 \pm 0.15$, $kT_e = 0.36 \pm 0.02$ keV and the other shows $\log\tau = 10.47 \pm 0.01$, $kT_e = 2.4 \pm 0.1$ keV, respectively. The major part of Fe-K line e blends come from the high T_e component which shows substantially higher line energy than that of the data. In spite of the good fit to the data, we should note that it is difficult to interpret the two-NEI-component model from the astrophysical point of view.

B Cassiopeia A

Cassiopeia-A is one of the brightest SNRs among those showing the shell structure. There are several emission lines from various metals as shown in figure 2. The spectrum and its restored image based on the observation during the PV phase was reported in Holt et al. [5]. There are three emission lines detected from Si and S: $K\alpha$ and $K\beta$ from the helium like ion and $K\alpha$ from the hydrogen like ion. They noticed that the line ratios among these three lines of Si were almost independent of position. Therefore, they concluded that the plasma condition of Si was independent of position. Whereas, they noticed that the line center energy of Si $K\alpha$ varied from position to position.

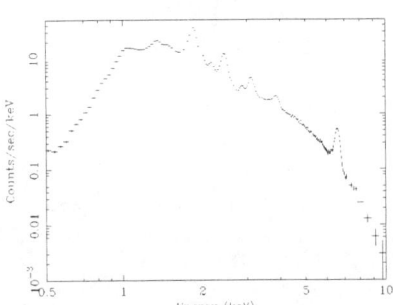

FIGURE 2. X-ray spectrum of the north west part of Cassiopeia-A with the *ASCA* SIS.

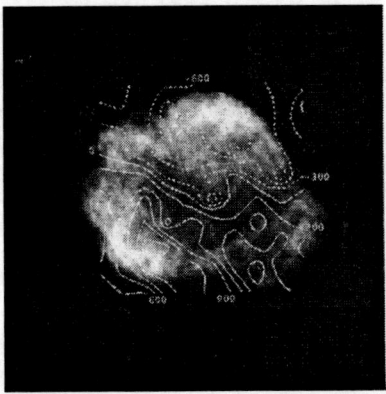

FIGURE 3. the Doppler shift velocity map of Si line superposed on the restored X-ray image of Si. The dashed lines show the red-shifted contour and the solid lines show the blue shifted contour. Numbers show the velocity in km/sec. The apparent size of Cassiopeia-A is about 4' in diameter.

Assuming that a simple NEI model of thin thermal emission [12,13], we derived

T_e and τ by the line intensity ratios for three lines of individual species. We obtained $kT_e= 1.80 \pm 0.07$ keV, $\log\tau = 11.2 \pm 0.2$ for Si and $kT_e= 1.85 \pm 0.09$ keV, $\log\tau = 11.3 \pm 0.2$ for S. The expected line center energies for Si and S $K\alpha$ lines observed by *ASCA* are almost constant since the ionization states of Si and S are almost in He-like ion. We found that the uncertainty in plasma conditions for Si and S corresponded to the difference in the line center energies of them to be 1-2 eV. Whereas, we found much bigger differences in line center energies among them. We found that the energy difference of the line center energies of Si and S was different from location to location. The energy difference is about 20 eV between the emission lines in the north bright region and those in the south bright region. This value can be explained neither by the CCD performance nor by the difference of the plasma condition. The possible explanation is due to the Doppler shift of the dynamic motion of the hot gas which is proposed by Markert *et al.* [11] based on the FPCS observation on board *Einstein*. We also conclude that the major part of the differences in line center energies of Si and S comes from the Doppler shift due to the dynamic motion of the plasma.

We divided the data into 10×10 small circles of diameter of $1'$ separating by $0.5'$. Each small circle is overlapping by $0.5'$ each other. The data from each small circle are analyzed in order to obtain the line center energies. In this way, we obtained line center energies for prominent $K\alpha$ emission lines for various elements. In this way, we measured relative line center energy map for Si, S, Ar, Ca and Fe which can be interpreted as the Doppler velocity. We showed the Doppler shift maps of Si and S superposed on the restored images in figure 3. We should note that the Doppler velocity map for S is qualitatively identical to that of S.

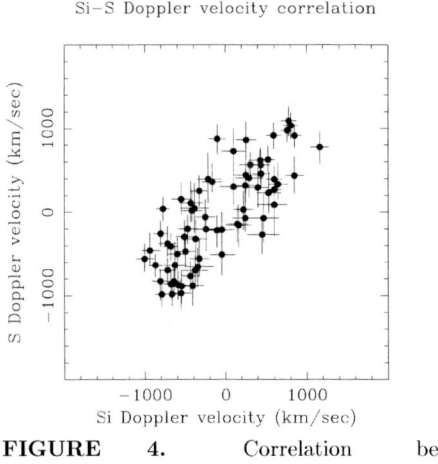

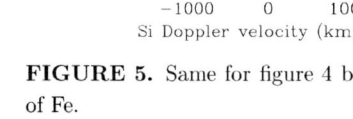

FIGURE 4. Correlation between the Doppler velocity of Si and that of S for various locations in Cassiopeia-A.

FIGURE 5. Same for figure 4 but of Si and of Fe.

Figure 4 shows the correlation between the Doppler velocity of Si, V_{Si}, that of

S, V_S. Each data point shows the result from each small circle. There is a clear correlation between them. We found $V_S/V_{Si} = 1.15 \pm 0.08$. The velocity difference inside Cassiopeia-A is about 2000km/sec. The Doppler velocity map in figure 3 shows that the north part is running away and the south part is running towards us. Furthermore, it shows not in the point symmetric but in the axial symmetric. If the explosion occurs in the point symmetric, any physical parameter including the Doppler shift map is also represented in the point symmetric shape. Since the Doppler shift map shows an axial symmetry, the plasma motion also shows an axial symmetry. The plausible explanation is that the hot gas is in the torus expanding radially whose velocity is proportional to the radius. We can conclude that the hot S gas is expanding similar way to that of Si.

We also obtained the similar result for other elements. Assuming that the line center energy of Fe is also due to the Doppler motion, we calculated the correlation between the Doppler velocity of Fe, V_{Fe} and V_{Si} shown in figure 5. We notice that there is also a clear correlation between them whereas, we obtained $V_{Fe}/V_{Si} = 1.7 \pm 0.2$. We should note that the Doppler shift map of Fe is also qualitatively similar to that of Si. If the difference of the line center energy of Fe is not due to the dynamic motion but due to the local difference in the plasma condition, there must be a clear correlation between the plasma condition of Fe and the dynamic motion of Si. The latter depends on the relation between the source location and the observer while the former is independent of it. We believe it implausible to explain the line center energy of Fe by the difference in the ionization state.

In optical region, Kamper et al. [7] detected the fast moving knot (FMK) and the quasi-statically flocculi (QSF). FMK is moving up to 6000 km/sec while QSF is moving at most several hundred km/sec. FMK is believed to originate from the explosion debris of the SN while QSF is believed to originate from the circumstellar matter surrounding the progenitor star. If it is the case, we can expect that the X-ray emitting gas also consists of two components with different expanding velocity and with different metal abundance. Due to the insufficient spatial resolution and the spectral resolution, we can not distinguish the emission from the fast moving gas and that from the slow moving gas.

C SN1006

SN1006 is a typical shell type SNR as shown in figure 6. This is a very unique SNR since the X-ray spectrum from the shell region clearly shows the power law type spectrum while that from the interior shows a thin thermal emission. The *ASCA* SIS clearly revealed the difference between them as shown in figure 7. It is considered as the site where the high energy cosmic ray is generated at the shock front. This is believed to be a high energy cosmic ray source [9]. The second similar source, RX J1713.7-3946, is also confirmed by the *ASCA* SIS [10].

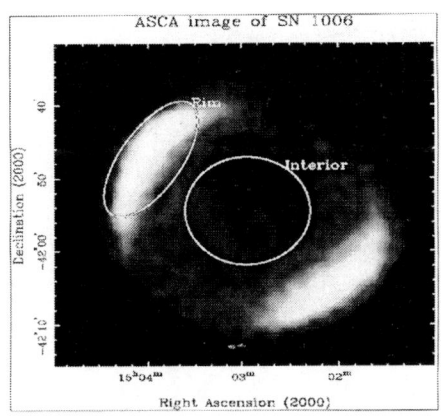
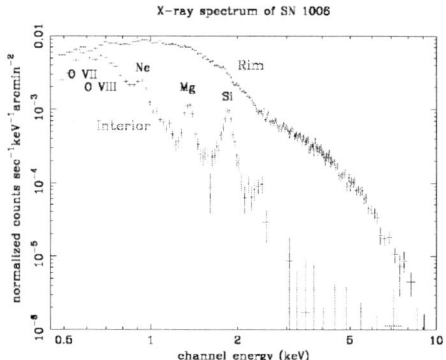

FIGURE 6. X-ray image of SN1006 observed with the *ASCA* GIS. The spectra for the Rim and that for the Interior are shown in figure 7. The apparent size of SN1006 is about 30′ in diameter.

FIGURE 7. The X-ray spectra obtained with the *ASCA* SIS for two regions shown in figure 6. The Rim shows the power law type spectrum while the Interior shows the thin thermal spectrum.

III PULSARS ASSOCIATED WITH THE SNR

The *ASCA* SIS has a good energy resolving power (FWHM at 6 keV is about 2%) whereas the time resolution is not so good. The time resolution of the SIS is $4 \sim 16$ second in the normal mode, depending on how many chips are used. Even in the timing mode, the time resolution is limited by the image size of the XRT. On the contrary, the *ASCA* GIS has a good time resolution with a moderate energy resolving power (FWHM at 6 keV is about 7%). The time resolution depends on the operation mode down to 61μsec.

The GIS is an imaging detector with high time resolution. The imaging capability improves the signal to noise ratio while the high time resolution covers a wide frequency range. It clearly shows that the GIS is good at detection of the X-ray pulsar. Torii *et al.* [18] developed a pulsar search algorithm: the Maximum Power Mapping (MPM) method. They divided FOV of the GIS into many small sectors and performed the timing analysis for each sector. They calculated the maximum power normalized by the expectation from the Poisson statistics. Mapping consists of the maximum power for each sector. Figure 8 shows how well does this algorithm work. The left figure shows the intensity map in the direction of AXJ1749.2-2725 which is heavily contaminated by a nearby strong source. The right figure shows the MPM which reveals the pulsation from AXJ1749.2-2725. This technique clearly reveals the association of the X-ray pulsar with the young SNR.

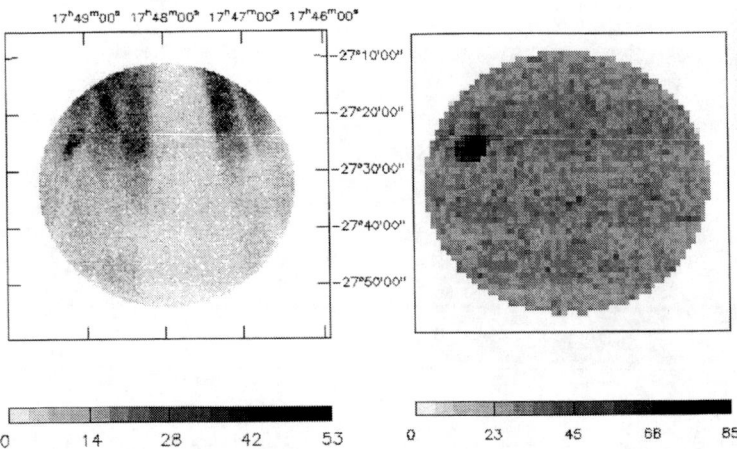

FIGURE 8. Left: the X-ray intensity map in the direction of AXJ1749.2-2725. The field is seriously suffered by the stray light from the bright X-ray source outside the field of view. Right: the Maximum Power Mapping clearly resolved the pulsar which is hard to be seen in the right figure.

A   G11.2-0.3

The source of G11.2-0.3 is considered as a historical remnant occurred in AD386 [15]. This is well known as a shell like SNR which is shown in figure 9 obtained with *ROSAT* [14]. The soft X-ray image (effective energy range is up to 2 keV) was found to be closely correspond with the radio image.

ASCA observed this source in 1994 during the AO1 program. The low energy map (0.5–3 keV) in figure 10-left shows a similar structure to that in figure 9 with taking into account the difference in the detection efficiency as well as the spatial resolution. Whereas, the high energy map (3–10 keV) shown in figure 10-right shows a center filled structure suggesting a hard source in its center. The X-ray spectrum of the center filled SNR is usually expressed with a combination of a single temperature bremsstrahlung, some Gaussians and a power law components. The bremsstrahlung and Gaussians represent the thermal emission from the shock front forming a shell structure while the power law component represents the synchrotron emission from the central source. Vasisht *et al.*(1996) [19] obtained $kT_e = 0.73^{+0.11}_{-0.08}$ keV with an interstellar absorption feature to be $1.38^{+0.08}_{-0.09} \times 10^{22} cm^{-2}$ and a photon index of $\gamma = 1.40^{+0.54}_{-0.68}$.

The existence of a hard plerionic component inside the SNR strongly suggests the presence of a rotation powered pulsar. Torii *et al.* [16] performed a careful timing analysis and found a clear coherent pulsation from the center. As shown in figure 11, a clear pulsation is seen at the frequency 15.463837 ± 0.000008.

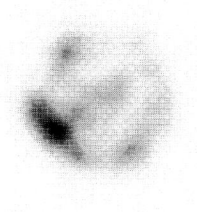
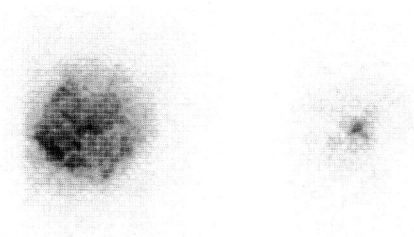

FIGURE 9. The intensity map of G11.2-0.3 obtained with *ROSAT*. The apparent size is about 4′ in diameter (HEASARC Research Center Online Service).

FIGURE 10. Left: the intensity map of G11.2-0.3 obtained with the *ASCA* SIS in the energy region of 0.5–3 keV. Right: same to the left but in the energy region of 3–10 keV. Only the central source is seen.

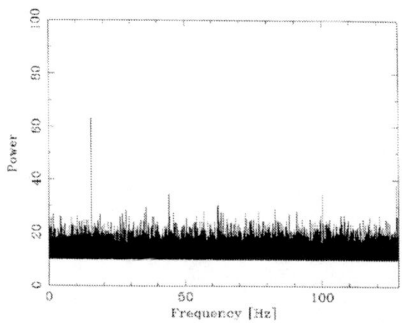

FIGURE 11. The power spectrum of the G11.2-0.3 data in the energy region of 3–10 keV. A clear pulsation at 15.463837 Hz is seen.

B RCW103

RCW103 is also a typical shell-like SNR as shown in figure 12 with a compact source (1E 161348-5055) in its center. Therefore, many people expected the compact source in the center a neutron star formed by the SN explosion. Aoki *et al.* [1] reported a possible detection of coherent pulsation of 69.319 msec from RCW103 with the *GINGA* satellite. Since the *GINGA* satellite had a F.O.V. of 1° × 2°, they failed to locate the pulsation.

Figure 13 shows an intensity map of RCW103 both for soft X-ray energy band (0.5–3 keV) and for hard X-ray energy band (3–10 keV). In the soft band image surely shows an extended structure which is similar to that obtained with *ROSAT*. However, there are two distinct sources in the hard band image. The source in the

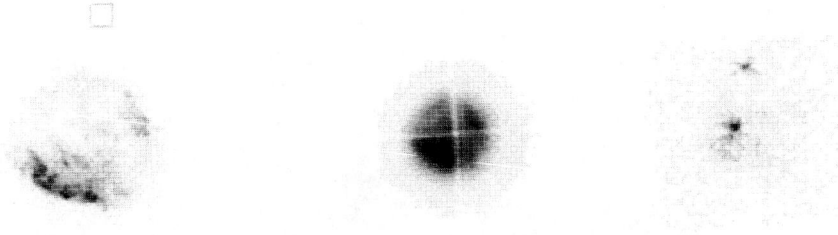

FIGURE 12. The intensity map of RCW103 obtained with *ROSAT*(HEASARC Research Center Online Service). There is no source in the upper square where *ASCA* discovered a hard source.

FIGURE 13. Left: the intensity map of RCW103 obtained with the *ASCA* SIS in the energy region of 0.5–3 keV. Right: same to the left but in the energy region of 3–10 keV. The upper source is in the square in figure 12. The angular distance between these two sources is about 8′. The scale of this map is different from that of figure 12.

lower part is 1E 161348-5055 and that in the upper part is a newly discovered hard source, AXS J161730-505505 [3].

Torii et al. [17] searched the entire field of view and found the coherent pulsation not from 1E 161348-5055 but from AXS J161730-505505. The detected period is 69.33802 ± 0.00003 msec. The source of AXS J161730-505505 is quite hard source which is not seen in the *ROSAT* data in figure 12 where the square represents the source location.

Due to the relatively strong source contamination from the nebula of RCW103, it is difficult to extract the precise source spectrum. We found that the center source, 1E 161348-5055, is soft with no pulsation and that the northern source, AXS J161730-505505, is hard with pulsation. Based on the combined analysis between the *ASCA* data and the *GINGA* data, we obtained the characteristic age of the pulsar to be 8000 years showing that the kick velocity of the pulsar to be faster than 820 km/sec. It is not clear whether AXS J161730-505505 is physically associated with RCW103 or not. If it is the case, the proper motion of AXS J161730-505505 will be 52 miliarcseconds/year which may be detected in near future.

IV SUMMARY

We presented here a very limited number of topics on the supernova remnant. *ASCA* has observed several tens of SNR and related sources. The *ASCA* SIS obtained X-ray spectra with good energy resolution. With the combination of the X-ray telescopes, the effective energy range expands up to 10 keV where we can cover most of the metal emissions. If the source shows thermal emission, many emission lines are clearly resolved. Young SNRs are surely in the NEI condition rather than the CIE condition. Based on this knowledge, we can determine the metal abundance precisely and the type of the SNR.

The *ASCA* GIS has a large field of view and a high time resolution down to 61μsec, depending on its operation mode. The GIS has a high performance in the pulsar detection. In particular, when the pulsar associated with the SNR is strongly contaminated by the nebulosity in the low energy range, its hard spectrum makes it prominent in the high energy range. The imaging capability of the GIS improves the S/N very much which is demonstrated by the MPM method.

REFERENCES

1. Aoki, T., Dotani, T., & Mitsuda, K (1992) *IAU Circ.*, 5588
2. Aschenbach, B., 1996, MPE Report, 263, 213
3. Gotthelf, E. V. and Vasisht, G., (1997) *Astropys. J. Letters*, **486**, 133.
4. Green, D. A., 1996, http://www.mrao.cam.ac.uk/surveys/snrs/
5. Holt, S. S., Gotthelf, E. V., Tsunemi, H. and Negoro, H. (1994) *Publ. Astron. Soc. Japan*, **46**, L151
6. Hwang U., Gotthelf E.V., 1997, *Astrophy. J.*, **475**, 665
7. Kamper, K. and van den Bergh, S. (1976) *Astrophy. J. Suppl.*, **32**, 351
8. Kinugasa and Tsunemi, H., (1999) *Publ. Astron. Soc. Japan*, **51**, 239.
9. Koyama, K., Petre, R., Gotthelf, E. V., Hwang, U., Matsuura, M., Ozaki, M. and Holt, S. S., (1995) *Nature*, **378**, 255
10. Koyama, K., Kinugasa, K., Matsuzaki, K., Nishiuchi, M., Sugizaki, M., Torii, K., Yamauchi, S. and Aschenbach, B., (1997), *Publ. Astron. Soc. Japan*, **49**, L7
11. Markert, T. H., Canizares, C. R., Clark, G. W. and Winkler, P. F. (1983) *Astrophy. J.*, **268**, 134
12. Masai, K. (1984) *Astrophy. and Space Science*, **98**, 367
13. Masai, K. (1994) *Astrophy. J.* , **437**, 770
14. Reynolds, S. P., Lyutikov, M., Blandford, R. D., & Sewards, F. D., *MNRAS*, **271**, L1
15. Stephenson, F. R. & Clark, D. H., (1978), *Application of Early Astronomical Records*, Adam Hilger Ltd., Bristol
16. Torii, K., Tsunemi, H., Dotani, T. and Mitsuda, K. (1997) *Astrophy. J. Letters*, **489**, L145.
17. Torii, K., Kinugasa, K., Toneri, T., *et al.*, (1998) *Astrophy. J. Letters*, **494**, L207.
18. Torii, K., Kinugasa, K., Katayama, K. *et al.*, (1998) *Astrophy. J.*, **508**, 854.
19. Vasisht, G., Aoki, T, Dotani, T., Kulkarni, S. R. & Nagase, F., (1996) *Astrophy. J. Letter*, **456**, 59.

CATSAT: A Small Satellite for Studying Gamma-Ray Bursts

W.T. Vestrand*[†], D.J. Forrest*, K.A. Levenson*, C. Whitford[‡],
D. Fletcher-Holmes[‡], A. Wells[‡], and A. Owens**

Space Science Center, University of New Hampshire, Durham, NH 03824
[‡]*University of Leicester, Leicester LE1 7RH, UK*
[†]*NIS-2, Los Alamos National Laboratory, Los Alamos, NM 87545*
**ESA/ESTEC, 2200 AG Noordwijk, The Netherlands*

Abstract. The Cooperative Astrophysics and Technology Satellite (CATSAT) is a University Explorer (UNEX) Class Mission that is being constructed by the University of New Hampshire and the University of Leicester. The primary scientific goal of the mission is to study the spectral properties of gamma-ray bursts in the energy range range from 500 eV to 5 MeV with particular emphasis on the 500 eV to 10 keV energy band. The satellite will be zenith pointed and flown in a 590 km sun-synchronous terminator orbit. Here we briefly discuss the mission and the expected scientific results.

INTRODUCTION

Scientists at public and private Universities have long been interested in space missions for Energetic Astrophysics. In fact, many of the early instruments carried on NASA high-energy astrophysics missions evolved from instruments developed at Universities for high altitude balloon and rocket experiments. Those early experimental efforts always had a strong educational component and much of the work, including design, construction, and operation, was carried out by students. However as high energy astrophysics entered the "Great Observatory" Era, the instrumentation for space missions became large, complex, expensive, and was developed on timescales that precluded significant student involvement with the space hardware.

In an effort to reverse this drift away from "hands-on" student involvement with space hardware, the Universities Space Research Association (USRA) announced, in May 1994, the STudent Explorer Demonstration Initiative (STEDI) which was a "pilot program to assess the efficacy of smaller, low-cost space missions as the essential element of a civil space program that is matched to the traditional process of research and development at universities". The goal of the NASA sponsored program is to demonstrate that small, relatively low-cost, space missions can be designed, constructed, and operated in a University Environment and that those missions can both enrich education and produce significant science. The total cost for each selected mission, excluding launch costs, was not to exceed $4.3 million. The plan to use Ultralite Expendable Vehicles launched from Vandenberg Air Force Base imposed additional constraints on the original mission designs such as a low earth polar orbit with inclination from 90° to 110° and a total payload mass of less than ~150 kg depending on required orbital altitude.

Universities across the United States, showed considerable interest in the program and a total of 66 proposals were submitted in response to the first Announcement of Opportunity. Ultimately, three of the proposals were selected to become the first STEDI missions. Here we discuss one of those missions, the Cooperative Astrophysics and Technology Satellite (CATSAT), which is being constructed by students, faculty, and staff at the University of New Hampshire with assistance from colleagues at the University of Leicester in the United Kingdom.

SCIENTIFIC OBJECTIVES AND INSTRUMENTATION

One of the greatest enigmas that astrophysicists currently face is the nature and the origin of Gamma Ray Bursts (GRB). Recent studies of burst afterglows have led to a consensus that GRBs are generated at cosmological distances [1,2], however, the nature of GRB progenitors remains unclear. The progenitors must be capable of releasing an outburst of electromagnetic energy in a few seconds that is comparable to the annihilation of the total rest mass energy of the Sun and that exceeds the electromagnetic energy from a supernova by a factor of 100.

To explain the tremendous energy release from a compact progenitor, theorists have been forced to invoke exotic mechanisms that can be divided into two general classes: (a) mergers of compact objects in close binary systems and (b) the collapse of unusual massive stars [3]. An important distinction between these two classes is the expected nature of the local environment around the progenitor. The typical timescale for the merger of compact objects such as neutron stars in a close binary system is of the order of a billion years. Since the velocities of those systems are expected to be high, reaching up to 1,000 km sec^{-1}, they should usually occur in rarefied "clean" environments well away from the original star formation region. In contrast, the collapse of a massive star, such as that proposed in the Hypernova scenario, should occur on a timescale of a few million years and be located in or near the dense clouds that are active star formation regions.

Soft x-ray measurements of gamma-ray bursts can distinguish between the two types of progenitor environments. Photoelectric absorption of x-rays produces a distinctive soft x-ray spectral cutoff at an energy that scales with the column depth of material traversed by the radiation (Figure 1). The presence of absorbing material in the vicinity of the progenitor can then be determined by comparing column depth derived from soft x-ray measurements of the burster with the galactic line-of-sight column depth that is known with great precision from radio surveys.

To effectively measure photoelectric absorption in GRB spectra for likely column depths, a soft x-ray spectrometer must have a sub-keV threshold [4]. To date, the lowest thresholds acheived during the GRB proper are the ~3 keV thresholds acheived by the GINGA and WATCH detectors.

CATSAT was designed to build apon the discoveries of the GINGA burst experiments [5–7]. It carries a suite of three instruments that will simultaneously measure GRB spectra over the broad energy range from 500 eV to 5 MeV. CATSAT's wide field-of-view and extension of the low energy threshold downward by nearly an order of magnitude into the soft x-ray range for GRBs will allow it to perform important spectral measurements of the prompt emission from GRBs.

Soft X-Ray Spectrometer

The primary scientific instrument aboard CATSAT is the Soft X-Ray spectrometer (SXR) which is as a wide field-of-view spectrometer covering the 500 eV to ~15 keV energy range. The SXR is composed of 112 individual 1.7 cm² Avalanche Photodiodes (APD) that are deployed in seven modules (Figure 2). Each module holds 16 APDs that are jointly collimated to yield a field-of-view (FOV) of approximately 1 steradian. The pointing directions for the modules are offset so that together they view nearly 2π steradians.

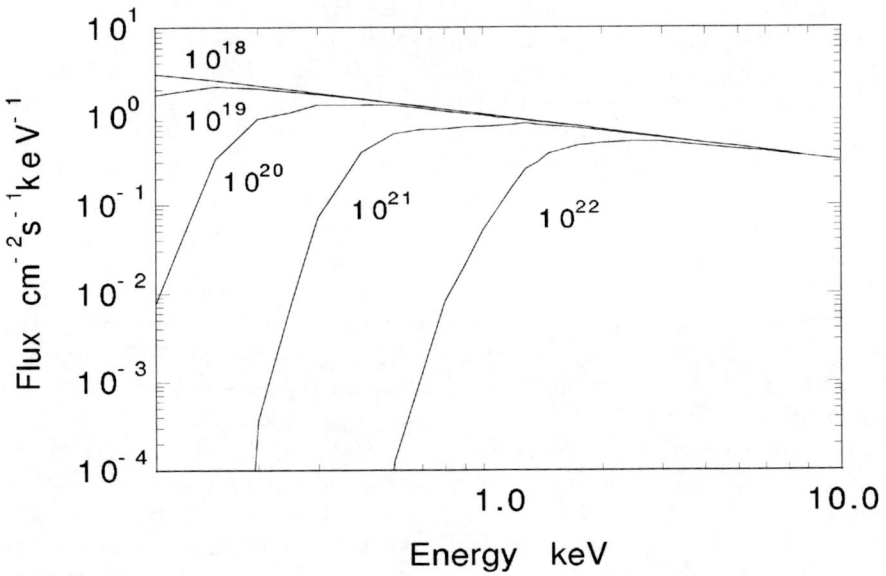

FIGURE 1. The soft x-ray spectra for a bright burst (with an intrinsic spectrum varying as $E^{-0.5}$) observed through column densisties of 10^{18-22} H cm^{-2}.

The use of APDs for direct soft x-ray detection is a new application for the devices, which are usually used to detect optical emission, and represents a key enabling technology for our low-cost mission. APDs are constructed from high purity n-type silicon wafers that are doped to form a p-n junction. A very high reverse bias ($\sim$1.5 KV) is applied across the diode so that charge carriers liberated by x-ray interactions in the silicon migrate into the depletion region, and are rapidly accelerated, generating a cascade of electron-hole pairs as they collide with silicon atoms. This charge multiplication makes APDs essentially a semiconductor analog of gas proportional counters. The high internal gain, in conjunction with modest cooling (-30°C) to reduce thermal noise, allows us to acheive energy thresholds below 500 eV for x-ray detection (Figure 3). To reduce the "noise" level generated by optical emission from bright stars and especially the moon, the APDs are covered by a thin Al filter.

The gain of an ADP is very sensitive to the bias voltage and temperature of the device. Therefore to control and match the gains of the APD detectors within an SXR module, CATSAT employs an Automatic Gain Control (AGC) system. The AGC system is composed of a high voltage power supply, Inflight Calibration Units (IFC), and circuitry that performs continuous active gain stabilization for each detector. The IFC units supply electronically tagged x-ray photons of known energy from ^{241}Am decay to the detectors for gain stabilization and calibration. The

FIGURE 2. The SXR assembly which is mounted on the zenith pointed side of the spacecraft. The insert shows a close-up of a SXR module with four collimators that each house 4 APD detectors and the central automatic gain control unit.

AGC circuits continuously screen event messages from the detectors and search for coincidence with an IFC signal. When a coincidence is detected, the measured pulse height is compared to the height expected for a calibration photon. The AGC circuitry then employs negative feedback to provide a small step-up or step-down signal to an optically isolated series high voltage regulator for that detector, thus stabilizing its energy response. The optical isolators also permit independent on/off control for each detector by command from the ground.

Hard X-Ray Spectrometer

The Hard X-Ray Spectrometer (HXR) functions as a wide field-of-view spectrometer covering the 15-380 keV energy range. It is composed of four independent $3" \times 0.4"$ $CaF_2(Eu)$ scintillation detectors. The four detectors are located on the corners of the SXR assembly and are canted at $45°$ so that altogether they observe 2π steradians of sky centered on the zenith direction. $CaF_2(Eu)$ has been selected as the scintillator material over NaI, even though it has a lower light yield per unit of energy deposition, because its low energy K-edge (<10 keV) provides a cleaner response in the 15 to 100 keV range. The straightforward response properties in that energy band and overlapping fields-of-view for the detectors should allow us to perform sensitive tests for the spectral features reported by the GINGA investigators [6].

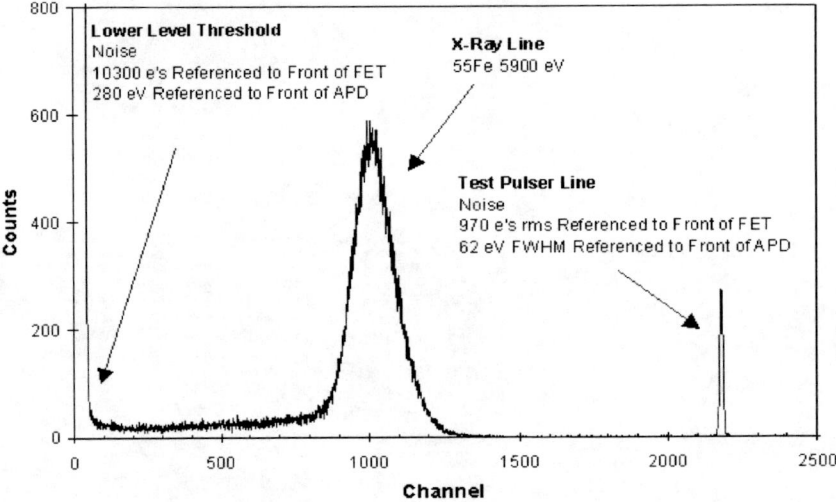

FIGURE 3. Laboratory measurements showing that an energy threshold below 500 eV is achievable with an ADP cooled to -30°C.

Gamma-Ray Spectrometer

The third spectroscopy system aboard CATSAT is the Directional Gamma Ray Spectrometer (DGS) which will measure burst spectra in the energy range from ~250 keV to 5 MeV. The DGS is composed of three NaI(Tl) scintillator/PMT assemblies arranged side-by-side in a close-packed array directly behind the SXR. Their collective angular response is nearly isotropic. However the close-packed arrangement of the three identical 3"×3" scintillation crystals causes a variable shielding of one another for most incidence directions. That variable shielding allows one to constrain the burst direction from the relative counting rates recorded by the three identical but independent detectors. The DGS direction constraints together with those provided the HXR and SXR systems will allow CATSAT to locate the burst direction to better than ~ 10°—an accuracy sufficient to allow precise spectral unfolding.

THE CATSAT SPACECRAFT

Before the CATSAT project, our experience with space missions at UNH and Leicester, which is typical for Universities involved in space research, had been

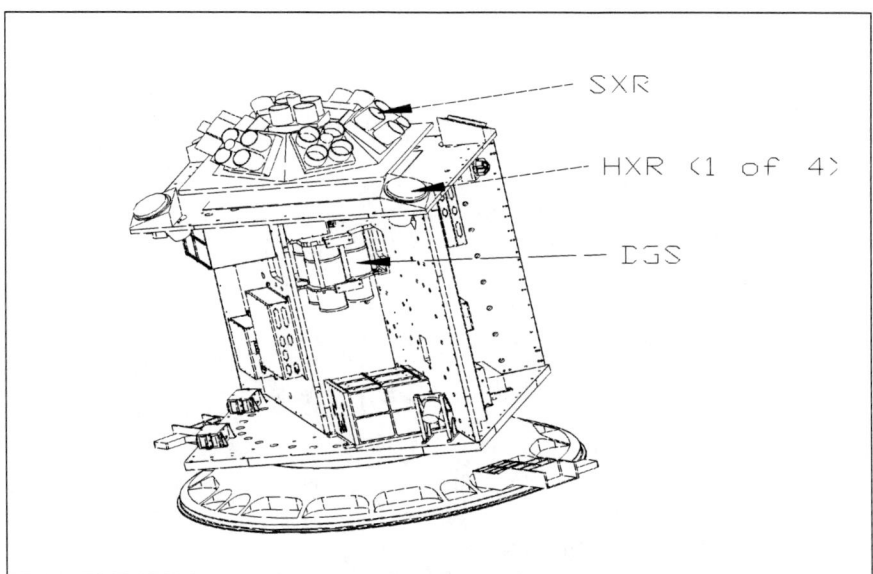

FIGURE 4. A cut-away drawing of the spacecraft frame. Here one of the vertical torque column panels at the left front has been removed to allow a view into the center of the column where the gamma-ray spectrometer is housed.

primarily limited to the construction of scientific instrumentation that was subsequently integrated with a spacecraft bus constructed by NASA, ESA, or one of the commercial Aerospace contactors. However, the strictures of budget forced important decision early in the design of the mission— Should we purchase a spacecraft bus from a commercial manufacturer and have a smaller fraction of the budget for scientific instrumentation or take the risk of building the bus within the university environment and use the money saved to construct instrumentation with greater scientific capabilities? We selected the latter approach.

Spacecraft Frame

The primary structural member of the spaceframe is a torque column formed from offset honeycomb panels. In cross-section the structure looks like a "pinwheel" with "blades" that extend from the central square tube to the corners of the rectangular box that forms the outer envelope of the frame. The panel extensions not only strengthen the structure but also allowed a flexibility of layout that permitted the payload components, which were being developed in parallel, to be distributed throughout the structure. A top deck which carries the SXR assembly and a bottom deck act as closing panels for the column (see figure 4).

Our NASTRAN modal and static load analysis indicated that standard 1/2" thick honeycomb panels for the top desk and vertical panels along with a 1" thick bottom deck would meet our requirements. To minimize expense, ready-made aluminum honeycomb panels with a standard 3/8" cell pattern were purchased from a commercial manufacturer. The panels were then milled at UNH to accept aluminum mounting spools and edge inserts. Additional cost savings were acheived by having UNH undergraduates perform the bulk of the insert potting with a cold cure epoxy.

In January 1998, the frame underwent vibratory testing and passed proto-flight qualification in all 3 axes. The 1st mode resonance of the frame was identified as 60 Hz lateral, well above the required 25 Hz minimum. The frame weight met our goal of less than 27 kg and the total expense for hardware and fabrication was only $75K.

On-Board Computer

CATSAT employs a single-board flight computer that was designed, built, and tested by students at UNH [8]. The flight computer is designed around a MIL-STD-883 Intel 80C186 microprocessor with a radiation tolerance of about 8 krads. That radiation tolerance, while modest for a spaceflight computer, is sufficient for a mission in low-earth orbit like CATSAT, and the 80C186 has already been used successfully on satellites constructed by the AMSAT community. The flight computer board also carries one Mbyte of EDAC protected SRAM, 64 kbytes of rad-hard PROM, a watchdog timer, and two RS-422 serial channels. All of the components have a radiation tolerance of at least 5 krad. Software for the computer is developed in C and can be debugged in real-time through an in-circuit 80186 emulator. At a production cost of under $20K, the CATSAT single-board computer bridges the gap between expensive commercial systems designed for more severe space environments and inexpensive commercial boards that are not designed to survive the temperature extremes and radiation dose found in space. Altogether, the CATSAT flight computer is a very capable and flexible design for low cost missions in low-earth orbit.

Attitude Control System

CATSAT will be launched into a sun-synchronous polar orbit with a 6 pm local time ascending node and an altitude of 590 km. The scientific operation of CATSAT requires pointing the x-axis (or roll axis) of the satellite, which is perpendicular to the solar panels on the sunward side of the satellite, to within 5° of the Sun. This requirement is placed by the need to shield the SXR, which is passively cooled to -40°C, from the Sun. The SXR, with its nearly 2π steradian field-of-view, must also be pointed near the local zenith to keep the Earth's limb out of the field-of-view. These pointing contraints dictate a 3-axis stabilized spacecraft—which is ambitious for a low-cost satellite.

CATSAT employs a biased momentum system for attitude control. The primary actuators are four reaction wheels symmetrically oriented in the X-Y plane and X-Z plane at an angle of 30° to the X-axis. This provides the capability for more momentum along the X-axis, which is the direction of the momentum bias, but also allows momentum components along the other axes that can be used for control. Additional control is provided by magnetic torque coils that are located in the X and Y planes and are split into two windings to provide redundancy.

Attitude determination is provided by measurements from an ensemble of on-board sensors. Two 3-axis magnetometers are employed: a commercial flux-gate magnetometer and an experimental magnetometer of our own design. There are

TABLE 1. Properties of the CATSAT mission.

Mission Parameter	CATSAT value
Physical Envelope	Fits within a 104 cm high cylinder with 101 cm diameter.
Mass	134 kg (295 lbs.)
Power	100 watts total. 60W for payload, 40W for bus.
Satellite Orbit	590 km Polar Orbit, Sun Synchronous terminator orbit.
Launcher	Delta II Rocket scheduled for July 2001.
Operational Lifetime	Greater than 1 year.
Attitude Control	Biased Momentum system with magnetic torque coils.
Attitude Determination	Better than 5° control with 1° knowledge.
Data Buffer	24 MB, holds 0.5 day continous data and 2 GRBs.
Communications	S-band transceivers at Durham,NH, USA and Leicester, UK.

coarse Sun sensors, each having FOV of $\sim 2\pi$, located on the +X and -X satellite faces and two additional fine Sun sensors on the +X face. Finally, there are two commerical infra-red horizon sensors located so that they view opposite limbs of the Earth.

Simulations of the attitude control system indicate that the combination of actuators, sensors, and control algorithms aboard the CATSAT satellite will provide the 5° control and the 1° pointing knowledge that are required to meet our scientific objectives [9]. Particular emphasis has been placed on simulations to throughly test the safe-hold procedure since it is the only attitude control software that is held in the spacecraft ROM and cannot be changed from the ground. That safehold mode, which will be activated during the initial spacecraft separation and whenever an attitude anomaly is detected, is designed to keep the solar panels pointing toward the Sun through a B-dot control algorithm.

CONCLUSION

The CATSAT program is on track for acheiving its goal of demonstrating that small satellite missions can be designed, constructed, and operated by students and faculty in a University environment. NASA's new committment to that model through the University Explorer (UNEX) program will provide students "hands-on" experience with flight hardware that has not been common since the early

FIGURE 5. A view of the sunward side of CATSAT showing the zenith pointing SXR, the deployed solar panels, and the sun-shield for the SXR.

days of the NASA spaceflight programs. Furthermore, CATSAT and the UNEX program should demonstrate that these relatively inexpensive missions can address important scientific questions and have a significant scientific impact that goes beyond the training that it provides students.

REFERENCES

1. Metzger, M.R., et al., *Nature* **387**, 878 (1997).
2. Kulkarni, S., et al., *Nature* **395**, 35 (1998).
3. Paczynski, B., *Astrophysical Journal* **501**, 467 (1998).
4. Owens, A., et al., *Astrophysical Journal* **447**, 279 (1995).
5. Murakami, T., et al., *Nature*, **335**, 234 (1988).
6. Fenimore, E., et al., *Astrophysical Journal* **335**, L71 (1988).
7. Strohmayer, T., et al., *Astrophysical Journal* **500**, 873 (1998).
8. Milani, D., *Proceedings of the 10th Annual Utah Small Satellite Conference*, (1996).
9. Whitford, C. and Forrest, D., *Proceedings of the 12th Annual Utah Small Satellite Conference*, (1998).